HISTORY, HISTORICITY AND SCIENCE

History, Historicity and Science

Edited by

TOM ROCKMORE
Duquesne University

JOSEPH MARGOLIS
Temple University

ASHGATE

Published by
Ashgate Publishing Limited
Gower House
Croft Road
Aldershot Burlington,
Hampshire GU11 3HR
England

Ashgate Publishing Company
Suite 420
101 Cherry Street
VT 05401-4405
USA

Ashgate website: http://www.ashgate.com

British Library Cataloguing in Publication Data
History, historicity and science
1.Science–Philosophy–History 2.Science and civilization
3.Historicism 4.Thought and thinking – History
I.Rockmore, Tom, 1942- II.Margolis, Joseph, 1924-
501

Library of Congress Cataloging-in-Publication Data
History, historicity and science / [edited by] Tom Rockmore and Joseph Margolis .
p. cm.
Includes bibliographical references and index.
ISBN 0-7546-5649-7 (hardback : alk. paper)
1. Science–Philosophy. 2. Historicism. 3. Science–History. I. Rockmore, Tom 1942- II. Margolis, Joseph 1924-

Q175.H4973 2006
501–dc22

2005037729

ISBN-13: 978-0-7546-5649-4
ISBN-10: 0-7546-5649-7

Printed and bound in Great Britain by MPG Books Ltd, Bodmin, Cornwall.

Contents

List of Contributors *vii*

Introduction 1
Tom Rockmore and Joseph Margolis

1 The Concepts of Physics: Rational Contents and Constructions in History 7
Michel Paty

2 Theory-change and the Logic of Enquiry: New Bearings in Philosophy of Science 31
Christopher Norris

3 Science, History and Philosophy in Kant and Hegel 77
Angelica Nuzzo

4 Historicity, Social Psychology and Change 95
Rom Harré and Fathali M. Moghaddam

5 The Reality of History 121
David Carr

6 The Social Location of Scientific Practices 137
Hugh Lacey

7 Kuhn, Different Worlds and Science as Historical 167
Tom Rockmore

8 Incommensurability Modestly Recovered 189
Joseph Margolis

Index *221*

List of Contributors

David Carr is Charles Howard Candler Professor of Philosophy at Emory University. He has worked extensively on phenomenology in particular. His publications include *The Paradox of Subjectivity* (1999), *Time, Narrative and History* (1986) and articles and essays on Husserl, phenomenology and the philosophy of history.

Rom Harré taught philosophy of science for many years at Oxford, where he is Emeritus Fellow of Linacre College. He has published widely on topics in the philosophy of physics and chemistry. Recently he has turned to studies in discursive psychology, a new development closely related to philosophy of mind. His most recent book, with Michael Tissaw, is *Wittgenstein and Psychology* (2005). Currently he teaches at Georgetown and American Universities in Washington, DC.

Hugh Lacey is Scheuer Family Professor of Philosophy Emeritus at Swarthmore College. He continues to teach at Universidade de São Paulo. His main interests lie in philosophy of science. He is the author of books in English and in Portugese, most recently *Is Science Value Free? Values and Scientific Understanding* (2005), which has also appeared in a Russian edition.

Joseph Margolis is Laura H. Carnell Professor of Philosophy at Temple University. His main interests lie in the philosophy of culture. He is the author of many books on different topics, most recently on moral philosophy after 9/11.

Fathali M. Moghaddam is Professor of Psychology at Georgetown University. His main academic efforts are toward developing psychology as a normative science. His most recent books are *Great Ideas in Psychology: A Cultural and Historical Introduction* (2005) and *From the Terrorists' Point of View* (2006).

Christopher Norris is Distinguished Research Professor in Philosophy at Cardiff University in Wales. He is the author of many books on philosophical topics, including (most recently) *Truth Matters: Realism, Anti-realism and*

Response-dependence (2002), *Hilary Putnam: Realism, Reason and the Uses of Uncertainty* (2002), *Philosophy of Language and the Challenge to Scientific Realism* (2004) and *Language, Logic and Epistemology: A Modal-realist Approach* (2004).

Angelica Nuzzo is Professor of Philosophy at Brooklyn College, CUNY. She has been Fellow at the Radcliffe Institute for Advanced Studies at Harvard (2000–2001) and is recipient of an Alexander V. Humboldt Fellowship (2005). She has written mainly on Kant, Hegel and German idealism in English, Italian and German. Her most recent book is *Kant and the Unity of Reason* (2005).

Michel Paty is Director of Research Emeritus at the Centre National de la Recherche Scientifique, and Invited Professor at the Philosophy Department of São Paulo University (Brazil). His research works deals in particular with the relationship between philosophy and the sciences during the classical (eighteenth century) and contemporary periods, with the mathematization of physics, with the questions of interpretation of quantum physics, and presently with the problems of a philosophy of scientific creation and of rationality. His most recent book is *La physique du XXe siècle* (2003).

Tom Rockmore is Professor of Philosophy at Duquesne University. His main interests lie in European philosophy, on which he has published widely. His most recent book is *Hegel, Idealism and Analytic Philosophy* (2005).

Introduction

Tom Rockmore and Joseph Margolis

To raise the question of how the very process of history affects the validity of science and our conception of knowledge sounds straightforward enough. But it is fraught with all the complexities spawned by the emergence of modern philosophy from, say, the middle of the seventeeth century to the beginning of the nineteenth, which, in effect, has indelibly colored the question down to our own day. There is no settled answer that any creditable part of the academy would be willing to endorse without deep misgivings; and yet it is hardly possible to raise the matter in a serious way without conceding that knowledge and thought are indeed inherently historical. We are caught in a naive paradox here: science, we are inclined to believe, seeks out the most far-reaching invariances that we can ascribe to the real world; and yet, we put all such discoveries at mortal risk by construing their achievement as the work of a contingent history that may well undo its own supposed success by a sudden turn in thought's own temporal life. Can that possibly make sense?

It would not be unreasonable to suppose that an untutored view of knowledge inclines toward a strong form of realism – meaning by that, that the independent world pretty well is as we perceive it to be: even a straight stick that looks bent in a glass of water, we learn to say, looks exactly as it should in such a circumstance. Our naive intuitions confirm, for one thing, that we do know the way the world is independently of our beliefs; for another, that, on the whole, our beliefs do not distort our perception of the real world, and, for a third, that what we thus come to know is not affected in any essential way by the admittedly historical process by which we achieve our level of scientific mastery. So the paradox dissolves, according to the strong realist doctrine. Nevertheless, the philosophical labors spanning the work of Descartes, Kant and Hegel, which still constitute ground zero for our understanding of the legitimacy of science itself, calls into question in the deepest way both the possibility of ever knowing the independent world as it truly is independently and the intelligibility of construing scientific knowledge as ahistorical.

Strange to say, this is still the stalemate of contemporary theories of science and history. There are almost no sustained analyses of the reconcilability of science's claim to objective standing and the historicity of thought and inquiry.

The essays we collect here address this striking and important issue in a fresh way. The truth is, it is a puzzle that first dawned on the Western world in the brief interval more or less confined to a sense of the dynamics of history anticipating and reflecting the events of the French Revolution. It gathered enormous conceptual force in the currents of philosophical speculation connecting Johann Herder's disappointed efforts to bring Kant's attention to what was put at risk in his great *Critique* by the complete absence of any reckoning of the significance of the history of thought and the encyclopedic labor, begun by Hegel in the *Phenomenology*, at the very start of the nineteenth century, in part responding to Herder and in part reconceiving Kant's project in the most uncompromising historicized terms. With Hegel, the entire question of the relationship between knowledge and history comes into full view armed with a sense of its likely resolution and of its unique role in marking the unity of the first four hundred years of modern philosophy.

That the issue of history and historicity has been largely neglected in the most ambitious efforts, in the twentieth century, to understand the essential epistemological puzzle posed by the splendid work of the physical sciences seems much more than shocking – seems well nigh unbelievable. In fact, what is perhaps the single most striking exception to the silence of the analytic tradition ranging over Anglo-American and Continental European philosophy – Thomas Kuhn's *Structure of Scientific Revolutions* (1962, 1970) – was itself systematically marginalized, almost completely eliminated from professional discussion, by the end of the twentieth century. The great enthusiasm for the philosophy of science launched in Berlin and Vienna before the full rise of Nazism, partly at least still under the influence of Kant, however qualified by Hegel's themes, that appear as well in Cassirer's *Philosophy of Symbolic Forms* and in Cassirer's subtle reflections on modern biology and physics, faltered rather badly over its own programmatic zeal. Much of late twentieth-century philosophy of science divided, self-consciously, between the continuation of the master projects of the reductive materialism and scientism of the first half of the century (though without much in the way of the fine-grained efforts of the early logical positivists and their closest allies: notably, among figures like Hans Reichenbach) and the relatively weak free-wheeling recovery of the problematic historicism of figures like Kuhn and Paul Feyerabend continued, for instance, in the study of actual laboratory life and the sociology of science suggested well before Kuhn by Ludwik Fleck.

The partisans of these two currents rarely addressed the same questions or attempted to resolve the questions that divided them. It was a great irony that Kuhn's enormously influential *Structure*, which had been well on its way to oblivion in the final decades of the twentieth century, should have been published in the *International Encyclopedia of Unified Science*, which was edited by Rudolf Carnap and Otto Neurath. Carnap was actually pleased with the book: he had himself turned away by that time from the extreme scientism of his earlier success. The contributors to the present volume write with a deep sense of the remarkable continuity of Western philosophies of science and knowledge and the problem of realism running like a red thread through the whole of its modern history – now stalled at almost the same impasse confronted in the late nineteenth and the whole of the twentieth centuries.

The idea of historicity is the double-edged sword of nineteenth- and twentieth-century Eurocentric philosophy. Seen against the backdrop of Kant's essential claims and the recuperative (but also corrective) import of Hegel's counter-labor, it consigns Kant's transcendentalism, together with Descartes's account of a 'strong realism,' to the deepest conceptual jeopardy. For its own part, historicity promises to replace, and render otiose, the supposed need for conceptual invariances and necessities of every substantive kind; but, in pursuing its own fortunes, it also faces reasonable questions about its own coherence and self-consistency. The classic contest that runs from Descartes to Hegel comes to rest in the seemingly unforeseen invention of the 'constructivist' solution to the Cartesian paradox. We must bear in mind that Hegel begins his *Phenomenology* with a very clever rehearsal of the *aporia* of the Cartesian approach to the problem of coming to know the independent world by way of internal mental representations of that world, which, nevertheless, unaccountably secure (for Descartes and Locke) the objective reliability of our sciences. The argument is as much addressed to Kant as it is to Descartes, inasmuch as Kant falls back to Cartesian representations (*Vorstellungen*) in spite of his having implicitly shown the way to defeating representationalism by constructivist means.

The peculiarity of Kant's solution of the Cartesian paradox rests with the Fact that Kant conjoins transcendentalism and constructivism indissolubly. Admitting the historicity of thought breaks the bond between the two doctrines: a historicized constructivism – let us say, Hegel's or any post-Hegelian alternative, possibly even Kuhn's, though Kuhn, of course, nowhere discusses Hegel's innovation – must unconditionally abandon transcendentalism, cognitive privilege, substantive necessities of any kind affecting the empirical work of the sciences. For example, the very idea of transcendental necessities may be converted, in historicized terms, into contingent, replaceable posits of what appear, as in the views of figures like Cassirer, William Whewell, C.I. Lewis or, more loosely, R.G. Collingwood, to be the best candidate transcendental constraints we can muster for organizing our epistemological vision as the necessary systematic ground of our actual sciences. There are bound to be many different strategies for reconciling science and historicity or, alternatively, for assessing the competing claims of transcendentalism and historicism. What is so remarkable in all this is that there seems to be no better grip on these enormously important problems – as much contemporary as early modern – than what a careful reading of Kant's and Hegel's original treatment of these issues would reveal. The fact is, the largest and best prospects for philosophy in our new century depend on a proper grasp of just this slim thread of conceptual history.

Eurocentric philosophy is now becalmed by its own sense of recycling more or less exhausted ways of addressing the persistent questions of our day without being able to put to rest the nagging puzzles of the Cartesian *aporia*, which should have been marked 'solved' a long time ago. We are in the middle of an uncertain search for a commanding vision apt to yield a measure of fresh success in every sector of philosophical interest. It is hardly by chance, therefore, that the general malaise affecting contemporary philosophy should accord with what answers to the profound stalemate at the heart of the philosophy of science. The contributors to the present volume are aware of this – aware, therefore, of the potential benefit of a close-grained analysis of the question before us. A pertinent discussion is likely to move through surprisingly diverse, possibly quite disparate, texts and topics; for, as we now understand matters, the issue posed by the historicity of science itself is, in its way, nothing less than a strategic glimpse of what looks to be the single executive question that, however much unrecognized, absorbs the whole of science and philosophy and cognitive inquiries of every kind.

Constructivism and historicity are, we are all inclined to wager, philosophy's best and amplest resource. It yields a general vision that is at least two hundred years old, but it remains largely unexplored. Apparently, conceptual change at its most global, or radical, moves with glacial majesty – which is to say, it hardly moves at all. Perhaps the conserving needs of language at its grandest theoretical sweep must pause long enough to prepare a number of enabling public idioms in advance of its eventual acceptance. That may explain a great deal of Kuhn's intuitive wisdom about paradigm shifts in the sciences, and may also, accordingly, bring Kuhn closer to Hegel's larger doctrine stirring once again – but in the archives. You must realize how radical a proposal it is to affirm, without ceremony, that constructivism and historicism – that is, the union of the two doctrines, not either one without the other – provide the best prospect for our philosophical future. You cannot think or say so, however, without rolling into a sentence or two a bold verdict on more than two thousand years of glorious history. And yet we must advance the charge, or fall back among our spent cartridges. That is to say, the short interval that spans Kant and Hegel was, philosophically, the most fateful – and prophetic – of the modern era, after its original self-discovery in the sixteenth century.

It's quite possible that adopting a constructive historicism may, in a single blow, put too much at risk to expect an easy transition from (say) a Cartesian conception to a Hegelian one. But the record shows, let it be acknowledged, that both Kant and Hegel were triumphant enough in their own day and that all the variant forms of Cartesianism were duly defeated by their respective demonstrations. We may imagine that there still remain options that we've overlooked. Why not? But that alone would hardly justify ignoring the known promise of the constructivist counter-strategy invented expressly to correct and replace the stubborn appeal of the naive vision that (even if defeated) still holds the academy in thrall. Extraordinary.

The point is not to hurry to formulate dismissals of an ill-remembered past. We would be well advised, rather, to lay out the arguments with due care, to connect them smartly with local histories – to match them not only with a patient recovery of their original historical force, but even more carefully, to spell them out in fresh ways fitted to every change in the contemporary sense of what the pertinent arguments really come to. The interval spanning Hegel and Kuhn should have taught us that a large part of the best work of philosophy lies with revivifying certain otherwise vestigial arguments through their various living counterparts, past or newly fashioned for a changing sensibility, that might catch up the *sittlich* details of the actual problems that they are meant to collect and resolve. The failure of the original sweep of Hegel's innovation to command philosophy's loyalty down to our own day signifies the academy's inertia and our need to retrace our steps again and again until the lesson is made secure. The rest is a matter of conviction and skill – on either side of every important contest.

Think of the challenge before us in the largest possible terms. Western philosophy, including the philosophy of science, has spent more than its first two thousand years committed to the idea that the independent structure of the world is changeless and yet also humanly and reliably discerned without inquiry's distorting the way the real world is. Now, think of this single vision as collecting endless variant forms through philosophy's entire trajectory down to our own day. There's a stunning summary for you – part history, part ideology. But imagine, further, that a fatal difficulty begins to make itself felt – say, through the gathering reflections of the seventeenth and eighteenth centuries – so that Kant suddenly springs into action, apparently from nowhere, hardly anticipated by his own speculations before the completion of the first *Critique*; he introduces a constructivist solution (that is, the inseparability of cognizer and cognized at the point of admitting the initial 'data' of any objective inquiry), but he leaves unchallenged the older notion of the invariant or necessary ground of any would-be cognitive order rightly fitted to the achievements of our science. Hence, his own invention yields an even deeper worry: he nowhere provides a convincing, uncompromisingly human ground for his own transcendental conjectures. For his part, Hegel, admiring Kant's immense labor, attempts in the most scrupulous way to reconcile Kant's solution of the paradox of knowledge with a dawning sense, informed by the perceived significance of the French Revolution, of the historically constituted, contingently evolving powers of human understanding. Hegel shows us, therefore, that the reflexive construction of a humanly contrived science (our science) must, in being historicized, be thoroughly contingent as well. So he shows us how historical contingency may take the form of apparent dialectical necessity. You see, therefore, how the entire run of canonical epistemological questions are suddenly put at risk. Historical regularities, subject to contingent and unforeseen change, threaten to displace changeless necessity itself – and thus challenge human regularities to recover a convincing sense of the evolving order of the experienced-world.

That is indeed the master theme of the deepest challenge to the Cartesian vision, which had itself originally appeared to weld the ancient and the modern world into a single continuous history. Subsequently, at the start of the nineteenth century, the old unity unravels, but is immediately reinterpreted along the lines of a supple historicism. Arguably, we ourselves are obliged to favor, in our own time, one variant or another of Hegel's admittedly florid revision of the canon. We cannot ignore its magisterial guidance, though we may perhaps escape its 'system.' Hegel's achievement decisively breaks the spell of philosophy's adherence to a single true thread of speculation running through its entire history. We cannot escape its lesson, though we may certainly misread it. At the very least, there is now an undeniable contest at the very heart of philosophy's history. It is a contest that demands that we reconsider what we should mean by truth and knowledge and meaning and reason and reality itself. These are the obvious topics that 'remain' – which is to say, the whole of philosophy itself. Well then, it comes as a surprise that we are only at the anticipated beginning of the investigation needed. The collection of essays offered here is intended as a reminder of unfinished business – and the need of a new beginning.

Chapter 1

The Concepts of Physics: Rational Contents and Constructions in History[1]

Michel Paty

1 Introduction. Objective Representation and Symbolic Thought

Physicists (like other scientists in a way) work with systems of laws and theories, with physical principles and concepts which are so well established that they appear untouchable and definitive: if the concepts on which they work are not completely secured yet, the very basis of such work is taken for granted. They thus hold, in their view, the true correspondence, in representational thought, between what exists in nature – bodies, radiations, objects, properties – and the entities referred to as 'physical systems' in general. Since these representations, or these elements of representation, are rational (they are related to each other and justified by reasoning, proofs, deductions, inferences and so on), this very rationality seems to be formulated as imaging objective elements, independent of thought, which are supposed to exist in nature: or if not imaging them exactly, such rationality is thought to be based on a narrow correspondence with them.

The fact that these objects or systems really exist (or that they can be produced) in nature is what gives them the character of being 'physical'. Properties are said to be 'physical' (one speaks of physical properties of physical systems) and so are also said to be the elements of knowledge of these properties (principles, concepts, quantities and so forth: one speaks of physical principles, physical concepts, physical quantities, and so on) that correspond to phenomena that have been characterized in nature. Anyway, such is the current conception that physicists have in mind when they ordinarily think in their work of the objects of their discipline. 'To think physically' is to think in terms of phenomena or objects of nature, whose understanding is referred to the physical sciences (in contrast to the biological sciences, for instance).

Physicists are fond of saying, about this stable knowledge on the basis of which they do their research, that phenomena or the things they describe are as

R. J. McEliece, *The Theory of Information of Information and Coding*, Addison-Wesley, 1977.

R. J. McEliece, *Finite Fields for Computer Scientists and Engineers*, Kluwer Academic Publishers, 1987.

W. W. Peterson and E. J. Weldon, Jr., *Error-Correcting Codes*, MIT Press, 1972.

V. Pless, *Introduction to the Theory of Error-Correcting Codes*, Wiley, 1982.

they describe them, and their discourse reflects this naive realism. According to them, their theoretical representations identify themselves with the 'objects' they account for: quarks, neutrons, elementary particles in general, atoms, black holes, quasars, galaxies. All these entities, infinitesimal or immense in size compared to us, that no one has strictly speaking ever seen directly, and that have become known to us indirectly (theoretically and experimentally or observationally), are supposed to exist firmly with the properties that their relative theories grant them; and not only to exist, but to be the very image of these descriptions: which, it should be mentioned, poses a problem (to them) when the very abstract and 'non-intuitive' character of these entities does not lend itself to visualization, which is certainly the case with the above-mentioned quantum particles, but also with black holes and quasars. Regarding galaxies, if we can have a visual image of them, in the distance, thanks to photographs from astronomical observatories, it is only by ignoring other aspects of their structure that are known otherwise than through sight (such as γ-ray bursts, non-traditional gravitational 'dark matter' and so on.). It is perhaps worth discussing here what 'existing' (as well as 'reality') means from the point of view of 'physics,' as well as the relation between what is said to 'exist' (or to 'be real') and its 'representation' by theoretical physics.[2] But we will not tackle these fundamental epistemological and philosophical questions directly here, although neither will we attempt to avoid them: they will be with us throughout our course, which will perhaps allow us to become a little more familiar with them.

However, be they physicists themselves, when their thought becomes reflexive, who take a step back and distance themselves from practice, or others who consider physics from the outside (such as philosophers, historians, sociologists or anthropologists of science, for instance), all know quite well that it cannot be a question of identification in a strict sense between this representation and the represented object, no matter how fine and powerful the physical theory. This 'object' supposedly belongs to nature and is outside of thought, whereas its 'representation' does not exist in nature, or exists only in a symbolic form, produced by human thought, conceived in subjectivity, received and communicated to others, shared and modified, through inter-subjective interaction and social relationships, but whose elementary instance is a knowing subject (rational or transcendental). Furthermore, this symbolic is built, under certain conditions that are those of thought (neurophysiological, psychological and subjective conditions), in a cultural (which includes its scientific and technical aspects), social, religious, ideological, political, economic environment, and in a way, it bears the signs of these conditions. These forms are not the same, depending on cultures and historical epochs.

This leads us to the problem of the subjectivity of knowledge, and of scientific knowledge in particular. Whether it is empirical or rational knowledge, it is true that it would be difficult to imagine knowledge without a subject to receive it, to assimilate it, to formulate it, or to renew it. Inasmuch as such knowledge corresponds to an 'integrated' view of elements, such as principles, laws, concepts and relations among concepts, such knowledge constitutes a corpus or a system (or perhaps several juxtaposed systems) which is rational to a certain degree. This is what is meant by the expression 'rational contents.'

By using this expression, I do not want to oppose the rational to the empirical, and I include for the moment in 'rational contents' the empirical that is aimed at knowledge, as long as it is rationally expressed: I will admit at this stage that rational contents may be heterogeneous, even if it means, later on, reconsidering the relationship for intelligible knowledge between the rational and the empirical. We know how much experience (in the sense of experimentation) is important to physics, to the point that physics is often said, by analytic philosophers, to be an 'empirical science.' This qualification does not appear to me to be fair, for it neglects one of physics's major characteristics, which is its privileged kinship to mathematics – and is there anything more eminently rational than mathematics?

It is true that physics cannot be confused with mathematics, since it relates to the real world, to the world of empirical phenomena. It, of course, speaks of the world (its phenomena, its objects) that it seeks to describe, but it does so in rational terms. At least it endeavors to do so, and it is only satisfied with its theories regarding physical phenomena when they are conceived as plainly rational, in their elements (principles and concepts) and in their structure (which connects the latter two). It can only achieve this by systematically and precisely using mathematics (this usage is not blind and obeys certain necessities, which can be characterized in each case, for each domain). To illustrate this better, we could mention the work of theoretical elaboration that led to the Newtonian and then the Laplacian theory of gravitation, to the analytic mechanics of Lagrange and Hamilton, to the mathematical theory of light since Fresnel, to the electromagnetic theory (Maxwell, Hertz, Lorentz), to thermodynamics (Thomson-Kelvin, Clausius, Boltzmann), to Poincaré's relativist electrodynamics, to Einstein's special and then general theory of relativity, to quantum theory, from quantum mechanics to the present theory of quantum fields and so on.

We have not exhausted, not even elucidated, by this assertion, the nature of the relationship between the empirical and the rational. At this initial stage of our reflection on the nature of concepts, I will limit myself to associating their objective aim (empirical phenomena) with their rational expression. I will mainly focus here on the apparent opposition between this character of concepts on the one hand, and the character of those same concepts of having been built through the unfolding of history on the other hand.

In this opposition – or better yet, in this tension – there is an instance, often forgotten when speaking of objective knowledge, because it seems to be opposed to objectivity, but without which the very word 'rational' would make no sense, and that is the subject of this knowledge, the basis of acts of judgment and reason. It is indeed through this instance of the subject that we established – and eventually solve – the tension between aimed objectivity and historical construction.

2 The Subject of Objective Knowledge

Even though the origin of the contents of knowledge is empirical, drawn from experimental and observational data, the contents of knowledge, that is to say the principles and the concepts, the relation of concepts and the laws, are expressed in rational terms (with several degrees of rationality, corresponding to greater or lesser transparency). We can consider them to be rational in so far that they are taken together in the unity and the coherence of a thought. Descartes was right in insisting on this capacity of subjectivity, 'equal in all men,' to organize rationally its knowledge and to make judgments, by doubting, by attaining certitude, or even by being hit (enlightened) by evidence, namely reason.[3] After him, and taking into account the critical contributions on the role of the senses and of empirical data (from Locke to Hume), Kant gave more precision to this foundation of knowledge by grounding it in a transcendental subject whose functions are such that this knowledge is made possible.[4]

This notion of the knowing subject was then criticized from very different fronts, but these criticisms did not eliminate in any way the necessity to invoke subjectivity as the place where the acts and the processes of knowledge occur, even if this subjectivity is seen as connected and (at least partially) determined by other instances, such as cognitive structures or social background.

The fact that knowledge is established in subjectivities is not *a priori* in opposition with the objectivity of scientific knowledge. The universality of reason invoked by Descartes, transcendental subjectivity in its formulation by Kant, and the conceptions proposed by the more recent important currents of philosophy of knowledge all aim at assuring or grounding the objectivity of scientific knowledge.

We will not here wonder about what is meant by 'objectivity,' whether it refers to a universal agreement among observers, or whether it is seen as correspondence between representational thought and the 'real world' (this latter expression designates nature as well as ideal forms like mathematics).

In fact, scientists feel in their work, or more generally in the thought of their science, the provisional and unfinished character of their even partial knowledge. They must sometimes, due to the very nature of their scientific endeavor, modify these elements (laws, theories, concepts, principles and so on) or replace them by others, which can be very different. They begin, in this case, by criticizing the ones that they have, showing their precariousness with the help of experimental data or of reasoning. Einstein, who spoke as an expert on the topic, wrote in 1935 in a text about 'Physics and Reality' that when the foundations of physics are questioned, as it was the case with quantum mechanics, 'the physicist cannot be satisfied with abandoning to philosophy the critical examination of the foundations of his science, since he knows best and senses where the shoe pinches. In his search for a new foundation, he must endeavour as much as possible to realize the pertinence, even the necessity, of the concepts that he uses'.[5]

Beyond the critique of the limitations of the previous knowledge, scientists can find themselves in the position of becoming a creator, discovering, inventing new elements of knowledge that they will substitute or that they will add to the former. The case of Einstein, with the two steps of the theory of relativity, is no doubt one that can speak best to a rather large number of people.[6] This attitude of creation is in fact more commonplace than imagined in physics (and in related fields such as chemistry, astronomy and astrophysics) as in the other sciences, and not only in the sciences of yesterday, but also in today's. Let us only consider a telling case: Pierre Curie, very well known for discovering, with Marie Curie, the properties of radioactive substances, established, many years earlier, his theory of symmetries in physics, which went almost unnoticed at the time, but became, later on, the gateway to an immense field in physical thought.[7]

In the critique which we have just spoken of, researchers (physicists, in this case) rejoin the judgments coming from the outside (by philosophers, historians, and so on) who reflect on the reasoning of scientific knowledge and on the contingency of the circumstances of its establishment. Both categories remember that the elements of knowledge, acquired and accepted, had not been given as eternally evident, but that they had been proposed at a moment of the history of the discipline in order to replace previous, older ones, and that these elements and their fitting were, in sum, constructions of the mind, conceived by the individual thoughts of this or that scientist, that they had been accepted by the scientific *milieu*, and that they had demonstrated their utility until they encountered their own limits. Here, the point of view of science (with physics as a particular case), which is that of 'internal necessity,' paying attention above all to the 'rational contents,' joins that of analysis and of reflections on science, which see them as 'constructions in history.'

In fact, through these preliminary considerations, we have already obtained a first result on the apparent opposition between rationality and objectivity of the contents of propositions of a science, such as physics, and the circumstances of their making, from which they have the status of historical constructs. This result is that in 'critical' moments (if not moments of 'crisis'), the two points of view coincide in the sense that conceptual or theoretical constructions (or reconstructions), historically situated, concern rational contents. If scientists are brought to reform, to reorganize or to transform their discipline, it is for reasons that concern these contents, which we have qualified as rational. We may already feel that, by taking into account the construction and the history, the rationality of the considered object is not destroyed, nor is it dissolved, but that it is situated.

We will now examine more carefully these conceptual contents of knowledge, what they mean, how they are constructed in history, and what happens in general to our way of conceiving the intelligibility and the rationality which are the last appeal to which the validity of these conceptual contents is referred. For if the history of concepts, of theories and of scientific knowledge in general is instructive in itself for learning about such elaborations, their meanings and their successions, it also provides us with precious information, for each step of the process, about what was meant then by 'intelligible' and 'rational.'

3 Conceptual Contents

Let us consider now the first term of the opposition or of the tension that we have previously mentioned: the 'conceptual contents.' To characterize conceptual contents, we must first re-state that scientific knowledge, like all human knowledge or belief, is expressed symbolically, in so far that it is an element of thought, thus substituting the forms of the real world, which we can only access through sensibility and understanding.[8] Symbols, in this sense, are not reduced to signs, even though they are ultimately presented as such, and physics certainly does not restrain itself due to its privileged use of mathematics.

A physical property is generally reduced to relations of magnitudes, which express (physical) concepts and are formulated in terms of mathematical quantities, as symbols (such as distances and coordinates of positions, durations and localizations in time, velocities, accelerations, energies and so on). The relations of concepts or physical magnitudes are then nothing but those of the corresponding mathematical quantities, subject (in so far as they are physical) to constraints like those asserted by the principles (such as the principles of mechanics, summarized by the principle of least action, which leads to equations of motions, or the principles of relativity, special or general and so on). The justification stands on the very constitution of the physical theory (as a structured ensemble of concepts represented as magnitudes mathematically expressed and governed by physical principles), and in the definition of the physical quantities as adequate to the description of the considered phenomena.

The system of physical concepts is woven by the mathematization of the magnitudes through which these concepts are expressed. In the end, it is this system that endows the physical content with these concepts and relations of concepts. I refer the reader here, for example, to the analysis of the physical meaning of the Lorentz transformation formulas as Einstein demonstrated them in his work of 1905 on special relativity, and to its consequences for the new conceptions of space and time. The latter's physical contents came directly from these transformations, deduced from the starting physical principles applied to the magnitudes of space and time taken respectively in the systems of reference in relative motion.[9]

Physical properties, however, are not just represented by magnitudes. They can also be represented by propositions, either by particular ones (from a result of a given experience) or general ones, such as, for instance, the physical principles, of relativity, of conservation, of symmetry and so on, which correspond to imposing very constraining conditions on the relations among magnitudes. The relations among spatial or spatio-temporal magnitudes can be metric or even topological. It is not necessarily the case, either, that the mathematical form of magnitudes adequate to the description of these physical phenomena and to the physical systems associated to these phenomena (whether they are their source or their place) must be a numerical function, even though this is their most common form, and that it has the advantage of being, possibly, directly put in connection with the indications of measuring instruments.[10] What is important in the relations between magnitudes or quantities expressing laws is the relation itself, more than the particular values attributed to these quantities or to their constituting elements.[11]

Let me make several remarks about this topic. First, generality is expressed through abstraction and, in a way, the range of the physical principles (which statement goes well beyond the particular facts that are at the origin of their formulation) has effects on the abstract character of concepts (rendered by mathematic symbolization) that they rule. The principle properties (which are statements of facts generalized in reason and are brought up to the rank of physical principles) can be expressed in relations of symbols. For instance, a principle of relativity or of symmetry is expressed as invariance for laws, or by a transformation of quantities that allows for this invariance (one speaks then of covariance).

Equations are thus obtained, which are symbolic constructs corresponding to effects (physical phenomena), and bearing consequences, as other effects, corresponding to predictions (on phenomena, properties of systems) or to previsions (trajectories and so on). With the principle of relativity restricted to motions of inertia, that is to say rectilinear or uniform, one obtains the invariant elements of the spatio-temporal metric (for four dimensions space–time), the mass–energy relation, the dilatation of durations and contraction of lengths, the invariance of equations of electromagnetism, and so forth. With general relativity, the red-shift of light frequencies, the slowing down of clocks (dilatation of duration), the curvature of space in the vicinity of great masses, and other effects multiplied in astrophysics and cosmology. The principles of invariance or of symmetry in actual quantum theories of fields also lead to explaining the organization of atomic and subatomic matter and enable the production and the identification of new phenomena.[12]

One can easily see from this that a symbolic physical representation (a representative theory, with its principles and its concepts) is not confined inside its symbolic world in the thought of the subject, 'in his brain': it can, by its effects, reach the external world. In this sense, scientific knowledge, even though it is constituted in the singularity of a subjectivity, corresponds to some objectivity: it transcends the subjective and even the inter-subjective and reaches the external world, the 'real' world, the 'physical world.'

If concepts, being transformed in the new theory that structures them, receive their physical meaning from the relations of the theory (and, ultimately, from the principles that rule those relations), and even if these relations are expressed by equations, it does not mean that the theory is reduced to the equations which express it, contrary to Heinrich Hertz's affirmation that 'the theory of Maxwell are the equations of Maxwell.'[13] Whereas equations are nothing but expressions of laws, the theory is more than the law it covers. The terms of these equations must, in order to be meaningful, be defined and conceived. The equations of Maxwell express the mutual variations of variables whose physical meaning is fixed only by relating these variables to physical entities defined as electric and magnetic fields, themselves affected beforehand by a certain physical meaning, given in the phenomena.[14]

To put it in a different way, even if the meaning content of the concepts is given in the relation, it cannot be reduced to the latter. Even though physical concepts, as magnitudes, are represented under the form of a mathematical expression and potentially bear the content of their mutual relations, their meaning contents do not have the simple mathematical transparency of the equations.

They are, in fact, far from this transparency for several reasons. On the one hand, they bear the potentiality of the mentioned effects (on the external world), which is due to their character as physical concepts. On the other hand, they are connected to propositions that are not expressed by magnitudes, yet they correspond to conditions on magnitudes, such as the above mentioned physical principles. It is essentially these principles that structure the physical theory, as we saw, and that ultimately determine the physical content of the concepts. They are, originally, facts of the empirical world, asserted and generalized. They still behold from this some obscurity, even if their formulation gets nearer to a mathematical expression (as with the condition of covariance of general relativity, or the principles of symmetry of groups of invariance in quantum theory of fields). Finally, we cannot neglect the fact that these concepts, these principles, these symbolic representations, are, and remain, in relation to a human activity; they constitute 'thought' and require, to be assimilated into thought, certain conditions that we shall call conditions of intelligibility.

Elements of symbolic representation, such as those in physics (and in science in general), thus possess two dimensions. One, directed toward the external world, corresponds to the physical content, and the other, internal, corresponds to integration in human thought. It is these two dimensions that one must hold accountable for the connection between subjective thought (or to be more precise, the thought of the knowing subject) and the physical world. We also see that the symbolic in science is not reducible to simple signs, since it bears a content that is oriented by two sides – the side of the world and the side of the mind.

These contents of representation in physics (and in other sciences), being irreducible to total transparency, because of this double dimension, are only intelligible to us through a synthetic grasp of thought, which is nothing other than what we usually call 'intuition.' Einstein expressed something quite similar by declaring to a correspondent: 'I believe that, ultimately, it goes the same with symbols in theoretical physics as with words in everyday language; all that is conceptual comes from the intuition.'[15] This means that we know and understand a phenomenon as a sort of synthesis, which presents itself to us, from an ensemble of relational properties which appear as immediately apprehensible, like a kind of 'second perception,' not through the senses, but through understanding: a kind of intellectual 'perception' that would give us the phenomenon by making it intelligible to us at the same time.

Here, we should discuss more carefully what I mean by 'intuition.' The meaning attributed to this word changes from one author to another, but one fact remains permanent: practically all authors, scientists and philosophers who have reflected about science and knowledge, from Aristotle to Descartes, Kant, Poincaré, Bergson, Einstein and others (with the exception of those who wanted to restrict themselves to a logical approach, from Leibniz to Russell and analytic philosophers) have felt the need, at one time or another, to appeal to the idea of intuition or to use this word to express something that they have experienced, even though they couldn't analyze it. It is this very same element of (individual) experience that we find here regarding intelligibility, and that, so to say, 'humanizes' the question of objectivity in science. An objective representation that would not be an intelligible representation would make no sense.

In other words, the question of objectivity in science presupposes that of intelligibility, generally forgotten, notably when the preoccupation is restricted to questions of justification and of logical foundation of science.

The question of the symbolic would also require discussion: it is in general the object of two mutually exclusive approaches: on the one hand the one that insists on the formal aspect of the symbolic brought back to signs (from Leibniz to Frege, Russell, Carnap), and on the other hand the one that is interested in the content of the symbolic thought of the transcendental or rational subject, from Descartes to Kant, Husserl, Cassirer and so on. It would be beneficial to take both into consideration. We can also add to that the importance of taking into account the idea defended by Poincaré and by Einstein: the 'free' relation (free 'in the logical sense,' or flexibility) of the thought of the knowing subject and the data on which he works (whether they are symbolic forms like the kind in mathematics, or relative to the empirical world given in perception and in experience, and known through the operations of understanding).

When we speak of scientific knowledge as being of symbolic character, we mean that this symbolic is not enclosed in itself, since it is capable of effects in the external world ('nature'). And if it is capable of such effects, it is not because it corresponds to some specie of supra-natural (as with shamanic representations, which are also symbolic human productions), but because it deliberately adjusts itself to nature, as much because of its conception of what phenomena are and of its recourse to experimentation as by the type of rationality that it admits. For sure, the common ground to all symbolic representations is their link to human thought and activity in general.[16] An element of symbolic representation corresponds to a content. The symbols are in thought, the effects in the world. The symbols–effects relation seems to concern two types of completely different entities. However, it is the contents that make the connection between both worlds, since they are directed toward the world through an idea of it. Action generated by thought occurs in the world.

In the case of a science like physics, the contents of the elements of symbolic thought are explicitly rational and are turned toward phenomena, toward the empirical world. Regarding knowledge in physics, what makes it specific among the sciences is the close relationship to mathematics (or to mathematization) as well as to experiment and quantitative observation. Already in place at the beginning of modern science, this double particularity assured the long supremacy of physics over the other branches of knowledge by letting it be considered, after mathematics (but the latter is not directed to the real world), as a model of scientific rationality.

4 Historical Construction

Let us now consider the expression 'historical construction,' the second element of the tension we analyze (between rational content and historical construction). We shall begin with some considerations on the meaning of the word 'historical' when speaking of the elements of a science, namely its concepts, that are historically constructed. Physics did not always exist as we now know it. Nature and natural objects, for instance, were represented differently at the time of Galileo or during the eighteenth century than at the turn of the ninteenth century. These representations were considered adequate, even true, in their time. If they were to be used today, they would be considered insufficient, approximate, outdated, and even partly false.

It is here that considering the time of history, for the present comprehension of what is science, becomes necessary. The two pitfalls would be, first, anachronism (to judge past knowledge through present values), and second, 'relativism' (to believe all knowledge to have equal value). We must therefore locate this scientific knowledge in its place and in its own time, and in the temporal succession that links the one to the others. Present knowledge has been made possible by past knowledge (though in an unpredictable way in general). We can follow these evolutions through the history of science, and recognize the moment when a new conception or view appears, the moment when it is received, again the moment when it, by being explicitly taken into account, changes the previous system of representation and so on. Science, the scientific theories as well the organization of disciplines and knowledge are not static, and evolve inter-dependently with a more global system of representations and culture (which includes technics, economics, social bounds) characteristic of an epoch, which is not necessarily a rigid system: in fact, there are multiple variants, forming, in a given context, a family, endowed with an overall characteristic.[17]

The evolution of the sciences points to their historical character. But to say that is not enough. History is not chronology. It is, as Marc Bloch used to say, the 'science of men in time,' and it is woven with mental representations, ideas, social life, esthetic and technical realizations, and so on. Scientific elaborations belong to the material and cultural history and of men in society. This requires us now to look more closely at the word 'construction' in the expression 'historical construction.'

Scientific concepts, concepts in physics in particular, but also principles and theories too, are historical constructions. So are the more general conceptions such as the very definition of physics, which used to refer to nature, but now designates one science of nature among others, having a precise goal, distinguished from others. Or the relationship between physics and mathematics, which became very narrow since the seventeenth and above all the eighteenth century, and which has become an indispensable constituting bond today due to the very necessity of thought, beacause of the modes of expression of magnitudes or quantities (geometrical, analytical, differential, partial derivatives, functions of abstract mathematical spaces and so on). So is the conception of a theory as a system of principles and concepts whose structure can be mathematically rendered (see the variation calculus, the principle of least action, the groups of transformation or of symmetry, the role of invariances and so on). So is the case with the modalities of experimentations, the techniques of observation, the role of instruments in the analysis and, more and more, in the production of phenomena, the relationship between theory and experience, conceptual abstraction and the results of measurement, the exigencies of adequation, of exactitude and so on. One can add, of course, the widening of physics, of its theories as well as of its objects, as opposed to reductionist attempts: mechanism, energetism, the 'electromagnetic viw of the world,' reduction to classical representations of observationalism for physics, to physico-chemistry for biology.

It would be possible to consider noteworthy cases in history of science that show the imbrication of all these elements at each step. For instance, the formulation by Galileo of the law of falling bodies represents an important moment of the constitution of physics in the modern sense, with the expression of a fundamental dynamic law. This moment can be characterized by a certain number of considerations whose ensemble determines, at the same time as a new knowledge (new contents of knowledge), a new manner of thinking (that we can relate to a configuration of rationality). The representation of physical concepts (for instance, velocity) by magnitudes or quantities means abandoning the qualities of scholastic thought; the correlative possibility of composing velocities among each other, even for different motions, violent (because of shocks) or continuous (such as free fall because by gravity), goes hand in hand with the idea that motion is no longer seen as inherent to bodies, the Galilean impulse (*impeto*) being seen as a communicated effect, and no more as the transmitted motor, as was the *impetus* of scholastic masters of fourteenth century (Jean Buridan, Robert Grosseteste, Nicole Oresme and so on).[18] Bodies are conceived independently of their state, whether in motion or at rest, and are not affected by motion (as Alexandre Koyré rightly pointed out[19]), and, motion and rest being ontologically equal, are from this perspective equivalent. From this the principles of inertia and of the relativity of motions result: to add or subtract velocity to the motion of a body does not affect its properties, that is to say, in fact, the laws of physics that are applied to it (equivalence of parabolic or vertical falls of objects, seen from points of reference in relative motion). Let us add to this the choice of 'sensible' magnitude (or concept) for the study of laws of motion, which is time, henceforth promoted to being the fundamental variable for physical laws.[20]

Or let us consider again the imbrications of conceptions in dynamics, geometry and differential calculus (or fluxions) in the works of Newton, Leibniz, and the brothers Jacques and Jean I Bernoulli,[21] and their disciples, then in the researches of Euler, Clairnaut and d'Alembert,[22] with their analytic study of properties of motions of solid bodies of any kind, fluids and continuous *milieux*, and objects of astronomy submitted to the action of several bodies. All these decisive innovations ended with the constitution of mechanics as analytic science (with Lagrange, Laplace and others[23]). That these elaborations represent more than a simple theoretical development in the line of a 'Newtonian paradigm,' dethroning a 'Cartesian paradigm' which was previously in place, becomes fairly clear when we examine the nature of the works of these researchers (who were then called the 'Géomètres'), notably those of the intermediary period, Euler, Clairaut and d'Alembert, who defined the new course of physics.[24]

Each one of them, in his way, was creative in a new domain and in an original direction, in transforming by the use of Leibnizian differential analysis the Newtonian formulation of the mechanics of bodies, taking from the Newtonian approach the concepts of impenetrability and attraction at a distance as well as the central idea of temporal variation. These concepts were foreign, no doubt, to Cartesian physics, but their way of considering the legitimacy of mathematically defining magnitudes was very different from the neo-platonic Newton's one, and in fact followed the exigencies of intelligibility expressed by Descartes in his *Rules for the Direction of the Mind* (*Règles pour la direction de l'esprit*) as well as his *Discourse on Method* (*Discours de la méthode*).[25] These exigencies could be filled, therefore justifying the mathematization of physics, thanks to the double function of mathematics in the exercise of reason according to Descartes (under the sign of *mathesis universalis*): serving as a model and as a guarantee of certitude in relations of propositions; ruling the expression of magnitudes appropriate to the representation of the physical world (continuous quantities subject to 'order and measure,' that is to say, submitted to quantitative relationships).

The epistemological implication of such a 'metaphysically' and 'ontologically' neutral conception of the relation of mathematics to nature was that the theories of physics can be mathematicized, because they put in place mathematically expressed magnitudes. Their mathematization was effective when these theories were in possession of magnitudes and of principles of physics appropriate to the considered objects and phenomena.[26] It was this conception that prevailed and was maintained through the use of new mathematical forms,[27] and the transformations of mathematical and theoretical physics,[28] and it was argumented by several mathematicians, physicists and philosophers, from d'Alembert to Kant, Ampère, Riemann, Boltzmann, Poincaré, Duhem, Hermann Weyl, Albert Einstein[29] and so forth.

So the mathematization of physical quantities took essentially, from the eighteenth century, the form of differential and integral calculus, appropriate to the treatment of continued magnitudes; and that the equations with partial derivatives, developed in conjunction with exigencies of fluid mechanics, became the 'language' of physics in continuous *milieux* and fields. These were at first thought through the 'substantial' support of a material *milieu*, such as phlogistic, caloric, ether and so on,[30] then simply as self-sufficient physical magnitude-concepts, such as combustion in presence of oxygen (with Lavoisier), fall of temperature between two sources of heat (from Sadi Carnot to Kelvin-Thomson and Clausius) or the propagation field with finite velocity (with Einstein's theory of relativity).

These magnitudes had, in classical physics and in the theory of relativity, the form of variables or the form of functions with numerical values, susceptible to being compared directly to the results of measurement. In quantum physics, magnitudes of this nature are the classical physical magnitudes, associated or 'projected' (quantum quantities over classical ones, with the corresponding probabilities), whereas the 'properly quantum' quantities are affected by more complex forms (such as the state functions, defined on mathematical Hilbert spaces, taking the form of linear superposition, and the operators, matrices, differentials, or others, acting on these functions). There exist still other types of magnitudes allowing for characterization of physical contents, such as strange attractors in the theory of dynamical systems, or topological non-metric magnitudes, that seem appropriate to some theories (for instance, in quantum cosmology).

It seems that what characterizes the physical range of the use of these magnitudes is above all their relational aspect, directly connected to phenomena. These magnitudes can be referred indirectly to classical magnitudes, if dealing with observation or measurement in the classical sense. But the specificity of physical phenomena is to be found at the level of relations of magnitudes proper to such systems: they can therefore be thought 'physically,' if we admit that what is 'physical' is more exactly defined by the fact of being theoretically significant (and corroborated by phenomena and experience, and so on), more than by the fact of being directly transcribed in terms of measurement.

The search of a direct correspondence of the *representation* with the *phenomena* that the representation aims at describing, which would characterize what is physical, would therefore have to focalize, on the one hand, on the *relationships of magnitudes* (be they at the start formal, and the idea of relationship finding its extension in *structure*) and, on the other hand, on the *specific characters* of these phenomena (which are the most interesting for physical thought, as they determine the new domain). Such a program for the definition of what is physical would indeed help in simplifying the (often unsolvable) problems of 'interpretation,'[31] to the cost of modifying the usual meaning of some categories of thought, such as the general notion of physical magnitude. Such changes of general conceptions and extensions of meaning for certain categories such as magnitude (for the notion of *physical magnitude* beyond the numerical and the measurable) correspond to modifications in the forms and the very modes of rational activity, to *transformations of rationality*.

5 The Transcendental Crucible or the *Rational* Thought of the *Empirical* World

As a matter of fact, how can we characterize the new conception of what is physical, as we have suggested above, if not by a modification of what is conceived as the very referent of comprehension? The case of magnitudes of quantum physics may appear to be specific, and it is also debatable since no general agreement exists (until now?) on the proposed extension of the notion of physical magnitude to the abstract concepts that were held for a long time to be 'purely mathematical,' as are state functions defined on Hilbert spaces and the corresponding operators acting on these functions.[32] Moreover, in this case, it would be nothing but a reinterpretation of an already acquired knowledge, which does not change the contents strictly speaking, and only proposes another way of seeing their meanings and of conceiving their intelligibility.

But intelligibility is not secondary, and brings in aspects of thought that go beyond mathematics and experience, and is concerned with metascientific and philosophical implications. Displaying differently the ordering of these diverse instances in the exercise of rational thought is all but innocuous, and it seems that it is always better to conceive their relationships as simply as possible, that is to say, to give the considered science the greatest autonomy *vis-à-vis* metascientific and philosophical questions. The rationality proper to physics gets more enlightened, and in this light we see it as transformed.

Furthermore, beyond its specificities, quantum physics represents only one particular case among other theories of physics and their histories: there are many examples of reinterpretation of knowledges, of experimentations and of formalisms, already acquired, that, when seen under new light, point to new referents for thought that speak directly for comprehension henceforth, restructuring what is called, for that science, 'intuition' as immediate synthetic knowledge. From this perspective, physical theory constitutes an instance of comprehension that no simple juxtaposition of mathematical relationships and of empirical data could equal.

By its very structure, connecting physical magnitudes regulated by principles that are also physical, under the species of equations, physical theory bears the physical content of its propositions, without having to superpose to it an interpretation, expressed in ordinary language, which would give meaning. Any true meaning assigned to it can be nothing else than these very relationships, which we, of course, can try to paraphrase or to comment upon in ordinary language, but which stand in themselves and that nothing can substitute. In the expression of these relationships, the form itself gives the meaning. To consider a few nowadays classical examples, how are we to imagine giving an account of laws of mechanics or of electromagnetism without using quantities expressed by differentials, thus provided, since their definition, with a physical sense, in so far as they are continuous? The undestanding of these domains of knowledge goes through the thought of magnitudes expressed in this manner.

In such cases, it is not so much a question of reinterpretation than the very process by which such knowledge is established. And, in physics, the *mathematical form* is inseparable from the thought of *physical contents*. It is clear, from Descartes to Riemann, Poincaré, Einstein and to today's physics, that the mathematical form is inherent to the thought of magnitudes, which no doubt gives physics an advantage over the other sciences of nature, considering its state of rationality, because the latter takes the simple form of the mathematical relationships between magnitudes. Unfortunately, this relative transparency presents some disadvantage: that of sometimes losing sight of the direct connection to the described reality (as it can be observed in the case of the current interpretation of quantum mechanics). Let me say here that the mathematization of physics, its privileged link to mathematics (which is a link of constitution at the level of the symbolic expression of concepts as magnitudes), is justified in so far as mathematics is adequate in expressing in the most tight-fitting and exact manner the relational character of magnitudes.

But physics, like other sciences, deals with the 'empirical' world given through experience, which it transforms inside the symbolic thought into intelligible rational propositions. In what way does this transformation of something empirical into something rational occur? Scientific thought constitutes, in this regard, a sort of crucible where this empirical matter is worked upon, in symbolic mode, and one would be tempted to say that it is a *transcendental crucible*, since the result of this work is rationally intelligible, beyond the data of perception and of experience. This transformation presupposes the formulation of principles of intelligibility, which are related expressly to the instance of the rational, under the aegis of which concepts are elaborated that come to impose themselves while remaining obscure in part, at least regarding an imaged representation (or a conceivable one in the exact terms of previously acquired representations). For instance, Newtonian attraction at a distance, with instantaneous propagation and without intermediary, or later on, the field, independent from any material or substantial support. These concepts, which have a *function of rational integration* of factual data, are thought of in relation to physical principles (conservation of energy, second principle of thermodynamic, principle of relativity and so forth, or 'formal' principles, least action principle, invariance principle, and so on), formulated from established properties, extended to the level of a universal validity, and so become the foundation of other knowledge of the considered domain.

Such notions or propositions invent, so to say, a new space for thought in physics that can be unfolded more widely, developing new relationships among its elements, which cannot be rendered by known terms, but on the contrary *bite* into a domain previously ignored, unknown, even unthinkable. They thus bring a wider and more comprehensive intelligibility capable of expanding the range of relationships beyond their old domain, hypostasizing the descriptive into a predictive.[33] These new propositions then *function* as rational principles, even though their reasons are not analyzable in terms of previous knowledge. What makes them admissible as necessary and rational is the *widening* of the domain and of the forms of *rationality*, capable of encompassing both new and old knowledges, in showing them from a same point of view and more deeply and thus conferring a better intelligibility.

In particular, to say of a magnitude that it is physical is to refer to it directly various elements of our knowledge of a domain, which no longer need any other intermediary to be justified. 'To understand otherwise,' more 'directly,' is to have 'practically realized' a modification in the very structure of thought itself, and more precisely in what we call 'rationality.' Such a widening of rationality for the sake of gaining a more immediate comprehension of phenomena or processes conceived until now only in an indirect and distant manner should not be exceptional since it is in this way only that the field of the unknown can open itself to intelligence.

The ways to penetrate what is still unknown are complex and require different procedures that lead to diverse degrees of knowledge before such knowledge is plainly grasped in understanding. We often *know* before we really *understand*: this knowledge is above all empirical or hypothetical, whereas understanding wants to be rational and requires consistency. The transformation occurs through the work of thought on symbolic elements that substitute the starting elements with rational terms (putting them in that way in relation). Physics gives us in this regard precise and exemplary, yet particular, indications, because the structures of rationality appear there to be narrowly connected to the mathematical form of relationships.

Notes

1 The ideas contained in this paper have been presented in French in a lecture for the cycle 'Thirty Years of Paris-7 University' (University Paris 7-Denis Diderot, 2 April 2001), and

in Portuguese as the inaugural lecture of the *Second Principia International Symposium*, Florianopolis (SC, Brésil), 6–10 August 2001. The French version is under publication in the *Proceedings* of this Symposium. The text has been translated into English by Farhang Erfani, as requested by the editors, and reviewed by the author.

2 See *La Matière dérobée* (Paty, 1988).

3 Descartes (1637).

4 Locke (1690), Hume (1748), Kant (1787).

5 Einstein (1935), in Einstein (1989–93), Vol. 5.

6 See Paty (1999c).

7 Pierre Curie, articles on symmetries, from 1884 and 1885, in Curie (1908), pp. 56–141. See Barbo (1999).

8 It is the very title of the classical work of Ernst Cassirer (1923–29), *The Philosophie of Symbolic Forms*. Regarding contemporary physics, see my *La matière dérobée*, which starts with this consideration (Paty, 1988).

9 Paty (1993), Ch. 4.

10 *A priori*, nothing obliges us to connect physical quantities to the possibility of measuring them directly, unless we are to subscribe to a certain view that makes of observation the primary reference of all knowledge, as does the 'observationalist' approach, which was that of the 'Copenhagen Interpretation' in quantum mechanics, promoted by Niels Bohr (Bohr, 1957, 1963).

11 I refer here to a remark made by Descartes in his *Rules for the direction of the mind* (*Règles sur la direction de l'esprit*) regarding the meaning of an algebraic expression (Rule 16), as well as to the examination to which he subjected the notion of magnitude subject to 'order and measure' (Rule 14), 'measure' meaning 'proportion': see Paty (2003).

12 Paty (1993a, 1988).

13 Hertz (1977).

14 In an advanced stage of the formulation of the theory, this physical meaning can be gathered from the enunciation of the principles, in the sense that we have discussed above.

15 Einstein, letter to H. Wolff, February 1949 (Archives Einstein); cited in Einstein (1989–93), Vol. 5, *Science, éthique, philosophie*, p. 127n. Regarding intuition in Einstein, see Paty (1993), Ch. 9.

16 Not being able to say more at this point regarding this question, I refer the reader to the perspectives opened, for instance, by Cassirer's *The Philosophy of Symbolic Forms* (1923–29).

17 We could discuss here the notion of paradigm proposed by Thomas Kuhn (Kuhn, 1960) that is too often used dogmatically and uncritically. Kuhn's idea seems too rigid and poorly defined, especially regarding the question of invention in science, and all things considered it is actually useless for precise historical analysis.

18 Crombie (1952), Clagett (1959).

19 Koyré (1935–39).

20 Galilée (1638). See Paty (1994b, 1996b).

21 Newton (1687, 1967–81); Leibniz (1849–1863, 1989), Fichant (1998); Bernoulli (1989–91), Blay (1992).

22 Euler (1911–), and the studies of Clifford A. Truesdell in the critical apparatus of the texts on mechanics in this edition; Clairaut (1743), Greenberg (1996), Passeron (1994); d'Alembert (1743, 1744, 1749, 1749–52, 1751–54), Hankins (1971), Paty (1977, 2001c and forthcoming), Emery and Monzani (1989), Grimberg (1998), Michel and Paty (2002).

23 Lagrange (1788), Galleto (1991); Laplace (1799–1825), Merleau-Ponty (1986), Martin-Viot (1994).

24 See, in particular, Paty (2001c, 2001d).
25 Descartes (1628), especially Rule 14; Descartes (1637). See Paty (1997a, 2003).
26 Paty (1994a).
27 From partial differential equations to complex numbers, to vectors, to tensors, to matrices, to spinors, to functions of 'integrable square' defined on Hilbert spaces and linear operators acting on these functions, to distributions, to probabilities, to non-commutative algebra, to topological notions, and so on.
28 Paty (1994a, 1999a).
29 Rieman (1854), Boltzmann (1974), Poincaré (1902, 1905, 1908, 1913), Duhem (1906), Weyl (1918), Einstein (1989–93), Vol. 5.
30 On the critique of the notion of substance and of the 'substantialist' thought, see: Cassirer (1910), Bachelard (1934, 1938).
31 On the 'orthodox' interpretation of quantum mechanics, see the particular philosophy developed by Niels Bohr (1957, 1963).
32 See Paty (1999d, 2000, 2003).
33 An example among others: the epistemological analysis done by d'Alembert, in the middle of the eighteenth century, of the Newtonian notions of attraction and impenetrability in order to find for them a justification in reason (see Paty, 2001b, pp. 82–3).

References

Bachelard, Gaston (1966 [1934]), *Le nouvel esprit scientifique*, Presses Universitaires de France, Paris.

Bachelard, Gaston (1986 [1938]), *La formation de l'esprit scientifique*, Paris.

Barbo, Loïc (1999), *Pierre Curie, 1859–1906. Le rêve scientifique*, Paris: Belin.

Bernoulli, Johann (1989–91), *Der Briefwechsel von Johann Bernoulli, 2: Der Briefwechsel mit Pierre Varignon, Erste Teil (1692–1702)*; *Zweiter Teil (1702–)*, Bearbeit und Kommentiert von Pierre Costabel und Jeanne Peiffer, Basle: Birkhauser (Vol. 1, 1988; Vol. 2, 1991).

Blay, Michel (1992), *La naissance de la mécanique analytique. La science du mouvement au tournant des XVIIè et XVIIIè siècles*, Paris: Presses Universitaires de France.

Bohr, Niels (1957), *Atomphysik og menneskelig erkendelse*, Kobenhavn. (German trans., *Atomphysik und menschliche Erkenntnis*, Wieveg, Brauschveig, 1958. English trans., *Atomic Physics and Human Knowledge*, New York: Wiley and Sons, 1958.)

Bohr, Niels (1963), *Essays 1958–1962 on Atomic Physics and Human Knowledge*, ed. Aage Bohr, London: Interscience.

Boltzmann, Ludwig (1974), *Theoretical Physics and Philosophical Problems, Selected Writings*, trans. from German by Paul Foulkes, Dordrecht: Reidel.

Cassirer, Ernst (1910), *Substanzbegriff und Funktionsbegriff*, Bruno Cassirer, Berlin. (Published as *Substance and Function*, in E. Cassirer, *Substance and Function and Einstein's Theory of Relativity*, English trans. by William Curtis Swabey and Mary Collins Swabey, Chicago: Open Court, 1923; New York: Dover, 1953.)

Cassirer, Ernst (1923–29), *Philosophie des symbolischen Formen*, 3 vols, 1923, 1925, 1929. (French trans. by Ole Hansen-Love and Jean, Lacoste, *La philosophie des formes symboliques*, Paris: Minuit, 3 vols, 1972.)

Clagett, Marshall (1959), *The Science of Mechanics in the Middle Ages*, Madison: University of Wisconsin Press.

Clairaut, Alexis (1743), *Traité de la forme de la Terre*, Paris.

Crombie, A.C. (1952), *Augustine to Galileo: The History of Science, A.D. 400–1650*, Falcon Press, London. (Re-ed. and augm., London: Heinemann, 1957.)
Curie, Pierre (1908), *Oeuvres*, Paris: Société Française de Physique, Gauthier-Villars.
D'Alembert, Jean le Rond (1743), *Traité de dynamique*, Paris: David (2nd edn, modif. and augm., Paris: David, 1758).
D'Alembert, Jean le Rond (1744), *Traité de l'équilibre et du mouvement des fluides*, Paris: David.
D'Alembert, Jean le Rond (1749), *Recherches sur la précession des équinoxes et sur la nutation de l'axe de la Terre dans le système newtonien*, Paris: David, 1749.
D'Alembert, Jean le Rond (1749–52), *Essai d'une nouvelle théorie de la résistance des fluides*, Paris: David, 1752, trans. d'Alembert from the original in Latin submitted to the concourse of the Academy of Berlin in November 1749.
D'Alembert, Jean le Rond (1754–56), *Recherches sur différents points importants du système du monde*, 3 vols, Paris.
Descartes, René (1628), *Regulæ ad directionem ingenii* (vers 1728), in René Descartes, *Oeuvres*, eds Paul Adam and Jules Tannery, Vrin, Paris (Re-ed., 1982, Vol. 10, pp. 349–488; French trans., *Règles pour la direction de l'esprit*, Paris: Vrin, 1970.)
Descartes, René (1637), *Discours de la méthode*, in René Descartes, *Oeuvres*, eds Paul Adam and Jules Tannery, Paris: Vrin, Vol. 6. (Re-ed., 1982, pp. 1–78.)
Duhem, Pierre (1906), *La théorie physique. Son objet, sa structure*. (2nd edn rev. and augm., 1914; re-ed., Paris: Vrin, 1981.)
Einstein, Albert (1989–93), *Oeuvres choisies*, ed. and trans. F. Balibar et al., Paris: Seuil, 6 vols, CNRS/Seuil, Vols 2 and 3, *Relativités* 1 and 2, 1993). (For the originals, see Einstein's Archives.)
Einstein, Albert, Lorentz, Hendryk, A., Minkowski, H. and Weyl, Hermann (1922), *Das Relaztivitätsprinzip*, with notes by Arnold Sommerfeld, Leipzig: Teubner. (English trans. by W. Perrett and G.B. Jeffrey, *The Principle of Relativity*, London: Methuen, 1923.)
Emery, Monique and Monzani, Pierre (eds) (1989), *Jean d'Alembert, savant et philosophe: portrait à plusieurs voix*, Paris: Archives contemporaines.
Euler, Leonhard (1911–), *Opera omnia*, Basel, 3 series of many volumes.
Fichant, Michel (1998), *Science et métaphysique dans Descartes et dans Leibniz*, Paris: Presses Universitaires de France.
Galilei, Galileo (1638), *Discorsi e dimostrazione matematiche intorno a due nuove scienze* (re-ed. with intro. and notes by A. Carugo and L. Geymonat, Boringhieri, 1958).
Galleto, D. (1991), 'Lagrange e le origini della *Mécanique Analytique*', *Giornale di fisica* 32, pp. 83–126.
Greenberg, John L. (1995), *The Problem of the Earth from Newton to Clairaut*, New York.
Grimberg, Gérard (1998), 'D'Alembert et les équations aux dérivées partielles en hydrodynamique', doctoral thesis, Université Paris 7-Denis Diderot.
Hankins, Thomas L. (1970), *Jean d'Alembert, Science and the Enlightenment*, Oxford: Oxford University Press.
Hertz, Heinrich Rudolf (1977), *Errinerungen. Briefe. Tagebücher. – Memoirs. Letters. Diaries*, arranged by Johanna Hertz. (Second enlarged edn prepared by Mathilde Hertz and Charles Süsskind, San Francisco, 1977.)
Hume, David (1748), *Philosophical Essays Concerning Human Understanding*. (Re-publ. under the title *Inquiry Concerning Human Understanding*, 1758.)
Kant, Immanuel (1787), *Kritik der reinen Vernunft*, Riga: J.F. Hartknoch. (2nd edn, 1787. French trans. by Alexandre J.L. Delamarre and François Marty, *Critique de la raison pure*, in Kant, Emmanuel, *Oeuvres philosophiques*, Vol. 1, Gallimard, Paris, 1980, pp. 705–1470.)

Koyré, Alexandre (1935–39), *Etudes galiléennes*, Paris, 3 vols. (2nd edn, 1 vol., 1966. (1935–39), Paris: Hermann, 1966.)

Kuhn, Thomas (1962), *The Structure of Scientific Revolutions.* (2nd enlarged edn, Chicago: University of Chicago Press, 1970.)

Lagrange, Joseph Louis (1788), *Mécanique analytique*, in Lagrange, *Oeuvres*, Vols 11 and 12, 1888 and 1889.

Laplace, Pierre Simon (1799–1825), *Traité de mécanique céleste*, 5 vols, Paris. (Re-ed. in 4 vols, Paris, 1829–1939.)

Leibniz, Wilhelm Gottfried (1849–1863), *Mathematische Schriften*, *Oeuvres*, ed. C.J. Gerhardt, Halle, 7 vols. (Re-ed., G. Olms, Hildesheim, 1962.)

Leibniz, Wilhelm Gottfried (1989), *Naissance du calcul différentiel, 26 articles des Acta Eruditorum*, intro., trans. and notes by Marc Parmentier, Paris: Vrins.

Locke, John (1690), *An Essay Concerning Human Understanding*, London.

Mach, Ernst (1883), *Die Mechanik.* (French trans. E. Bertrand, *La mécanique*, Hermann: Paris, 1904.)

Martin-Viot, Florence (1994), 'L'élaboration des principes variationnels en dynamique, de Lagrange à Hamilton et Jacobi', doctoral thesis, Université Paris 7-Denis Diderot.

Maxwell, James Clerk (1873), *A Treatise on Electricity and Magnetism.* (3rd edn (1891), 2 vols, re-ed., New York: Dover, 1954.)

Maxwell, James Clerk (1890), *The Scientific Papers of J.-C. M.*, Cambridge.

Minkowski, Hermann (1908), 'Raum und Zeit', *Physikalische Zeitschrift* 10, pp. 104–11. (English trans., 'Space and Time', in Einstein et al. (1923), pp. 73– 91.

Merleau-Ponty, Jacques (1983), *La science de l'univers à l'âge du positivisme. Etude sur les origines de la cosmologie contemporaine*, Paris: Vrin.

Michel, Alain and Paty, Michel (eds) (2001), *Analyse et dynamique. Etudes sur l'œuvre de d'Alembert*, Québec: Presses de l'Université Laval.

Newton, Isaac (1687), *Philosophiae Naturalis principia mathematica*, London. (2nd edn, 1713; 3rd edn, 1726, ed. with variants by Alexandre Koyré and I.B. Cohen, Cambridge, 1972. English trans. by A. Motte, *The Mathematical Principles of Natural Philosophy*, 1729; 3rd edn, 1726; trans. rev. by F. Cajori, Berkeley: University of California Press, 1934.)

Newton, Isaac (1967–81), *The Mathematical Papers of Sir I.N.*, 8 vols, ed. Derek T. Whiteside, Cambridge: Cambridge University Press.

Passeron, Irène (1994), 'Clairaut et la figure de la Terre au dix-huitième siècle. Cristallisation d'un nouveau style autour d'une pratique physico-mathématique', doctoral thesis, Université Paris 7-Denis Diderot.

Paty, Michel (1977), 'Théorie et pratique de la connaissance chez Jean d'Alembert', doctoral thesis, Université des Sciences Humaines, Strasbourg 2

Paty, Michel (1988), *La matière dérobée. L'appropriation critique de l'objet de la physique contemporaine*, Paris: Archives contemporaines.

Paty, Michel (1990), *L'analyse critique des sciences, ou le tétraèdre épistémologique (sciences, philosophie, épistémologie, histoire des sciences)*, Paris: L'Harmattan.

Paty, Michel (1992a), 'La dimension philosophique du travail scientifique (découverte et interprétation)', *Philosophique* (Université de Besançon), No. 1, pp. 45–67.

Paty, Michel (1992b), 'La question du statut de l'expérience en physique', *Cahiers d'Histoire et de philosophie des sciences*, No. 40, pp. 183–97. (Actes du Colloque *Les Procédures de preuve, de validation et d'évaluation dans les sciences et les techniques: une approche historique*, Lille, 11–13 April 1991.)

Paty, Michel (1993), *Einstein philosophe. La physique comme pratique philosophique*, Paris: Presses Universitaires de France.

Paty, Michel (1994a), 'Mesure, expérience et objet théorique en physique', in Jean-Claude Beaune (ed.), *La mesure, instruments et philosophies*, Champ Vallon, Seyssel (01), pp. 159–74.

Paty, Michel (1994b), 'Le caractère historique de l'adéquation des mathématiques à la physique', in Santiago Garma, Dominique Flament and Victor Navarro (eds), *Contra los titanes de la rutina. – Contre les titans de la routine*, Madrid: Comunidad de Madrid/C.S.I.C., pp. 401–28.

Paty, Michel (1994c), 'Sur l'histoire du problème du temps: le temps physique et les phénomènes', in Etienne Klein and Michel Spiro (eds), *Le temps et sa flèche*, Editions Frontières, Gif-sur-Yvette, pp. 21–58. (2nd edn, 1995; Collection Champs, Paris: Flammarion, 1996, pp. 21–58.)

Paty, Michel (1996a), 'Le style d'Einstein, la nature du travail scientifique et le problème de la découverte', *Revue philosophique de Louvain*, 94 (No. 3, August), pp. 447–70.

Paty, Michel (1996b), 'Galilée et la mathématisation du mouvement', *Passages*, No. 76, April–May, pp. 49–53.

Paty, Michel (1997), '"Mathesis universalis" e inteligibilidad en Descartes', Spanish trans. by Martha Cecilia Bustamente, in Victor R. Albis, Jorge Charum, Clara Helena Sanchez and Gonzalo Serrano (eds), *Memorias del Seminario en conmemoración de los 400 años del nacimiento de René Descartes*, Academia Colombiana de Ciencias Exactas, Físicas y Naturales, Coleccion Memorias, No. 9, Bogotá, pp. 135–70. (Portuguese trans. Maria Aparecida Corrêa-Paty, '"Mathesis universalis" e inteligibilidade em Descartes', *Cadernos de História e Filosofia da Ciência* (Campinas), Série 3, Vol. 8, 1998 (No. 1, January–June), pp. 9–57. Also published in *Seminário sobre O Cartesianismo*, Centro de Estudos de História e Filosofia da Ciência, Centros de Investigação da Universidade de Évora, Évora, Portugal, 2000, pp. 145–200. Published in the original French as '"Mathesis universalis" et intelligibilité chez Descartes', in Karine Chemla, Siegmund Probst, Agnès Erdély and Antonio Moretto (eds), *Ceci n'est pas un festschrift pour Imre Toth* (29.12.1996) 2004.)

Paty, Michel (1998a), 'La philosophie et la physique', Jean-François Mattéi (éd.), *Le Discours philosophique*, Vol. 4 of l'*Encyclopédie philosophique universelle*, Presses Universitaires de France, Paris, Ch. 123, pp. 2104–22.

Paty, Michel (1998b), 'Le vide matériel, ou: La matière crée l'espace', in Simon Diner and Edgard Gunzig (eds), *Univers du tout et du rien*, Bruxelles: Editions de l'Université de Bruxelles, pp. 22–44.

Paty, Michel (1998c), 'Les trois dimensions de l'espace et les quatre dimensions de l'espace-temps', in Dominique Flament (ed.), *Dimension, dimensions I*, Série Documents de travail, Fondation Maison des Sciences de l'Homme, Paris, pp. 87–112.

Paty, Michel (1999a), 'La place des principes dans la physique mathématique au sens de Poincaré', *Fundamenta philosophiæ* (Nancy/ed. Kimé, Paris) 3 (2), 1998–99, pp. 61–74.

Paty, Michel (1999b), 'Les trois stades du principe de relativité', *Les Relativités*, special edition of *Revue des questions scientifiques* (Namur, Be), 2000, pp. 103–50.

Paty, Michel (1999c), 'La création scientifique selon Poincaré et Einstein', in Michel Serfati (ed.), *La recherche de la vérité*, Coll. L'Écriture des Mathématiques, Paris: ACL-Editions du Kangourou, pp. 241–80.

Paty, Michel (1999d), 'Are Quantum Systems Physical Objects with Physical Properties?', *European Journal of Physics*, 20 (November), pp. 373–88.

Paty, Michel (2000), 'Interprétations et significations en physique quantique', *Revue Internationale de Philosophie* (Bruxelles), No. 212, 2 (June), pp. 199–242.

Paty, Michel (2001b), 'Intelligibilité et historicité (Science, rationalité, histoire)', in Juan José Saldaña (ed.), *Science and Cultural Diversity: Filling a Gap in the History of Science*, México: Cadernos de Quipu 5, pp. 59–95.

Paty, Michel (2001c), 'Les recherches actuelles sur d'Alembert. A propos de l'édition de ses *Oeuvres complètes*', in Alain Michel and Michel Paty (eds), *Analyse et dynamique. Etudes sur l'œuvre de d'Alembert*, Québec: Presses de l'Université Laval.

Paty, Michel (2001d), 'D'Alembert, la science newtonienne et l'héritage cartésien', *Corpus* (revue de philosophie, Corpus des œuvres de philosophie en langue française, Paris), No. 38: *D'Alembert* (ed. Francine Markovitz and Jean-Jacques Szczeciniarz), pp. 19–64.

Paty, Michel (2003), 'La notion de grandeur et la légitimité de la mathématisation en physique', in Miguel Espinoza (ed.), *Deuxième Journée de Philosophie des Sciences Jean Largeault*, Paris: L'Harmattan. (English version, 'The Idea of Quantity at the Origin of the Legitimacy of Mathematization in Physics', in Carol Gould (ed.), *Constructivism and Practice: Toward a Social and Historical Epistemology. A Volume Dedicated to Marx Wartofsky's Memory*, Lanham, MD: Rowman and Littlefield)

Poincaré, Henri (1902), *La science et l'hypothèse*, Paris: Flammarion (1968).

Poincaré, Henri (1905), *La valeur de la science*, Paris: Flammarion (1970).

Poincaré, Henri (1908), *Science et méthode*, Paris: Flammarion.

Poincaré, Henri (1913), *Dernières pensées*, Paris: Flammarion, 1913. (Re-ed. 1963.)

Riemann, Bernhard (1854), 'Ueber die Hypothesen, welche der Geometrie zugrunde liegen (Mémoire présenté le 10 juin 1854 à la Faculté philosophique de Göttingen)', *Abhandlungender königlischen Gessellschaft der Wissenschaften zu Göttingen*, Vol. 13, 1867. (Also published in B. Riemann (1902), *Gesammelte mathematische Werke. Nachträge*, ed. M. Noether and W. Wirtinger, Leipzig, pp. 272–87. French trans. by J. Houel, 'Sur les hypothèses qui servent de fondement à la géométrie', in B. Riemann, *Oeuvres mathématiques*, trans. L. Laugel, Paris, 1898; 1968.)

Weyl, Hermann (1918), *Raum, Zeit, Materie*, Berlin, 1918. (4th edn, augm., 1921. Trans. Gustave Juvet and Robert Leroy, *Temps, espace, matière. Leçons sur la théorie de la relativité générale*, Paris: Blanchard, 1922; Paris, 1979.)

Chapter 2

Theory-change and the Logic of Enquiry: New Bearings in Philosophy of Science

Christopher Norris

1

Anglo-American philosophy of science has tended to define itself squarely against the kinds of (so-called) 'metaphysical' approach that have characterized (so-called) 'Continental' philosophy in the line of descent from Husserl. Indeed, Husserl's project of phenomenological enquiry was the target of criticism by Frege – and later by Gilbert Ryle – which pretty much set the agenda for subsequent debate.[1] That project seemed to them just a form of ill-disguised 'psychologism,' one that purported to address issues of truth, validity, rational warrant, and so forth, but which fell far short of the logical rigour attained by thinkers in the other (analytic) tradition. Thus Husserl might claim – like Descartes and Kant before him – to be raising questions about the *a priori* forms of human knowledge and experience, forms that were given (necessarily presupposed) in every possible act of cognition.[2] Moreover, he might claim to have advanced beyond Kant in distinguishing more clearly between formal and transcendental logic, or judgments whose necessity followed from the ground rules of this or that logically binding system of thought, and judgments that resulted from a rigorous reflection on the genesis and structure of human understanding in general.[3] However these claims counted for little with Husserl's critics in the other, that is, post-Fregean 'analytical' camp. What they chiefly objected to in Husserl's project was the approach via thoughts and ideas 'in the mind' of some perceiving or reasoning subject, even though Husserl was very often at pains to reject any merely empirical (or psychologistic) construal of his claims. To their way of thinking, all this talk about 'transcendental' truth- and validity-conditions was just another variant of the bad old Cartesian-Kantian retreat to consciousness as the last court of appeal in epistemological matters. Only by rejecting that entire line of thought – that is to say, by adopting a strictly analytical or logico-semantic approach – could philosophy at last break free of its attachment to naive, subject-centered or 'metaphysical' notions of meaning and truth.

More recently some analytic philosophers, Michael Dummmett among them, have begun to question this doxastic account and to offer a more nuanced appraisal of the differences between Frege and Husserl.[4] Nevertheless there is still a marked tendency to suppose that, wherever such differences show up, it will always be a matter of Frege's having pointed the best (most rigorous and logically adequate) way forward, and of Husserl's having slipped – despite all his strenuous protests to the contrary – into some form of argument that unwittingly reveals his residual Cartesian premises. Now it seems to me that these belated, rather grudging gestures of atonement are partly a result of certain emergent problems within the analytic tradition, in particular those raised by Quine in his famous attack on the two 'last dogmas' of Carnap-style logical empiricism.[5] That is to say, they have to do with the impossibility – as Quine sees it – of maintaining any version of the logical-empiricist dichotomy between 'matters of fact' and 'truths of reason', or synthetic and analytic orders of judgment. This distinction was crucial to the logical-empiricist program, and also, in a somewhat less doctrinaire form, to the whole enterprise of analytical philosophy in the wake of Frege and Russell. Whether Quine actually succeeded in giving it the *coup de grâce* is still very much an open question, not least (as I shall argue) in view of his failure to offer any convincing alternative approach to issues of truth, meaning and interpretation.

So there is a case for taking a different view of the grounds for dispute between the two philosophical traditions. This revisionist approach would go much further than Dummett in acknowledging both the importance of Husserl's project and the various problems that analytic philosophy has encountered in its attempt to enforce the Fregean veto on talk of consciousness, intentionality, ideas, eidetic structures, the noetic/noematic distinction and so on. In so doing it would raise very different objections to logical empiricism from those put forward by Quine and other thinkers who continue to endorse that veto. After all, Quine is insistent that philosophy of language should adopt a strictly extensionalist semantics which finds no room for such 'opaque' (intensional) items as beliefs or propositional attitudes. Thus, for instance, while it is jointly true that 'W.V. Quine is a Harvard philosopher and author of the essay "Two Dogmas of Empiricism",' there is no equivalent conjunctive truth-value which attaches to that statement when preceded by the phrase 'Mary believes that ...'. For she might, as it happens, be in error about various facts, among them his academic calling, or his institutional affiliation, or his having written that particular essay, or perhaps about the *name* of whoever it was that answered to one or more of those identifying descriptions. (Maybe she thinks it was written by Hilary Putnam or some other well-known Harvard philosopher.) In short, such statements fail to meet the test of substitutivity *salva veritate* for terms – such as 'Quine' and 'the author of "Two Dogmas of Empiricism"' – which pick out an identical referent and are therefore extensionally equivalent. Hence also his argument against the possibility of quantifying into modal contexts, a practice which – on Quine's submission – creates all manner of insoluble problems, such as the breakdown of logical equivalence between '9 is necessarily greater than 7' and 'The number of planets is necessarily greater than 7.'[6]

Thus Quine has basically the same reason for objecting on the one hand to intensional (belief-related) approaches in philosophical semantics, and on the other to modal-logical talk of necessity, possibility, truth across various (more or less epistemically accessible) 'worlds,' and so forth. Where they both run into trouble is in failing to secure the functional equivalence of statements – those containing names or definite descriptions – whose truth-values *must* be covariant if we are to uphold (as Quine thinks we must) the extensionalist doctrine that terms with the self-same referent should always be interchangeable *salva veritate* in every context of usage. Nevertheless it is hard to see how this argument can possibly work given his relentless attack, in 'Two Dogmas', on the various co-implicated notions (of synonymy, analyticity, truth-by-definition, and so on) which in Quine's view reduce to so many items of otiose circular reasoning. Some would argue on the contrary that such reasoning can have a certain ampliative function, that is, a capacity for drawing out the sense and proper usage of those terms, despite their apparent circularity when treated from a strictly logical point of view.[7] Then again, other critics of Quine – Ruth Barcan Marcus among them – have argued that the problem lies not so much with quantifying into modal or intensional contexts, but rather with providing an adequate extensionalist theory of meaning and truth. On this view, Quine's standard objection (that modal/intensional approaches are 'referentially opaque') can in fact be turned around to claim that the opacity 'lies with Quine's use of such terms as "identity," "true identity," "equality"'.[8] Thus, according to Marcus, logical systems should be treated as 'more or less extensional,' the criteria in each specific case depending 'on the kinds of contexts and predicates which are prohibited, and the degree to which the relation of identity is equated to weaker forms of equivalence.'[9]

However, it is not my purpose here to adjudicate these issues within the broad spectrum of current analytical philosophy of language and logic. Rather, it is to make this more general point: that they all take rise from a certain preemptive narrowing of philosophic sights, namely the belief – one shared by Quine and Marcus – that we shall make progress or achieve more clarity in these matters only by eschewing the kinds of 'subjectivist' approach that are taken to characterize work in the other, that is, post-Husserlian Continental tradition. In what follows, I propose to question this belief and suggest that analytic philosophy has much to gain from a closer acquaintance with developments in Continental epistemology and philosophy of science. In particular I shall focus on two such developments, both carried on (albeit from a critical-deconstructive standpoint) in the early writings of Jacques Derrida. These are the projects of Husserlian phenomenology, and the distinctive approach to issues of scientific knowledge and theory-construction adopted by Gaston Bachelard. I shall argue that they offer a promising way forward from the various problems – especially those remarked upon by Quine – which have dogged the legacy of logical empiricism and its various successor-movements. From this point of view, the dispute between Quine and Marcus as to whether or not one can quantify into modal or inten*s*ional contexts is a regional dispute that has arisen in part as a consequence of their both foreclosing on that wider (inten*t*ional or consciousness-related) domain of enquiry.

'It seems to me,' Marcus writes, 'that much of the discussion these past few years concerning the apparent breakdowns of substitutivity principles in intensional contexts and its presumably devastating results for logic and mathematics are [*sic*] largely terminological.'[10] And again, more pointedly: 'the apparent difficulties of interpreting such [modal/intensional] systems are not genuine, but analogous to a rejection of non-Euclidean geometry because it allows parallel lines to meet.'[11] As it happens both Husserl and Bachelard are much concerned with the processes of thought by which advances come about in fields such as mathematics, geometry and physics, advances that can only be achieved – so they argue – through a break with hitherto accepted ideas of self-evident, 'commonsense' or *a priori* truth. What is at issue here – as Marcus brings out very clearly – is the question how far such processes of thought should figure in a properly logical account of the context of justification. For some analytic philosophers, Quine pre-eminent among them, there is just no place for inten*s*ionalist (let alone intentionalist) modes of talk when trying to make logical sense of such matters. For others – including Marcus – there is a firm line to be drawn between inten*s*ional approaches (which are strictly indispensable if one wishes to deploy modal, counterfactual, or other such causal-explanatory devices) and inten*t*ional approaches which supposedly open the way to all manner of psychologistic vagaries. Thus, on her account, logic can perfectly well accommodate most of the items that Quine objects to ('"knows that," "is aware that," and in particular "is necessary that"') without fear of referential opacity. However, it requires that those qualifying phrases be construed properly, that is to say, in accordance with a modal-logical or strictly intensionalist approach which finds no place for such inscrutable matters as the discovery-procedures or the processes of thought by which conjectural beliefs are transformed into adequate knowledge.

2

It is this dimension of enquiry that was firmly closed off by the 'analytic' turn against Husserlian phenomenology, and indeed, against all those later 'Continental' developments perceived as not matching up to the highest, most rigorous standards of logical accountability. My case is that Anglo-American philosophy was thereby deprived of some useful epistemo-critical resources, not least for overcoming the various problems bequeathed by logical empiricism, problems acknowledged – in their different ways – by Quine and Marcus. More specifically, the two projects of Husserl and Bachelard can be seen to address crucial issues that were mostly disregarded by thinkers in the mainstream analytical tradition, but which have often surfaced to vex that tradition in the form of such internal disputes as the rift over modal/intensional conceptions of meaning, logic and truth. Hence the lately emergent strain of 'post-analytic' philosophy, conceived – very often with reference to Quine – as an outcome of just these intractable problems in the doctrine of logical empiricism. I have argued elsewhere that this reactive movement of thought is frequently in danger of swinging right across to the opposite extreme, that is, toward a 'strong' hermeneutical approach (derived from Wittgenstein and Heidegger) or a full-blown Rortian 'textualist' position which finds no room for such old-fashioned notions as reality or truth.[12] Indeed, Quine's argument in 'Two Dogmas' is itself a very striking example of the way that a throroughly physicalist (or naturalized) epistemology can be made to square with an equally thoroughgoing doctrine of meaning-holism, one that takes the entirety of scientific knowledge or belief at any given term as the 'unit of empirical significance.' And from here it is no large step to those varieties of hermeneutic-textualist doctrine according to which interpretation goes 'all the way down,' so that any results turned up (say) through 'direct' observation or empirical research must themselves be treated as interpretive constructs, or as making sense only against some background 'horizon' of communal belief. For if indeed, as Quine writes, 'it is nonsense, and the root of much nonsense, to speak of a linguistic component and a factual component in the truth of any individual statement,' then equally it is the source of much confusion – or so

I shall argue – to push this case for meaning-holism to the point where 'truth' becomes entirely a matter of contextual definition.[13]

Thus for Quine, famously, knowledge is 'a man-made fabric which impinges on experience only along the edges,' or again, 'a field of force whose boundary conditions are experience.'[14] At the centre of the fabric are those so-called logical 'laws of thought' (bivalence, non-contradiction, excluded middle) which are taken to possess *a priori* validity or to function as the very ground-rules of rational debate in the physical sciences and elsewhere. However, Quine cautions, we should not be misled into thinking that these rules are *absolutely or intrinsically* immune from revision. For this is, after all, just another doomed attempt to uphold the artificial dichotomy – whether Humean, Leibnizian, Kantian, Carnapian or whatever – between contingent 'matters of fact' on the one hand and necessary 'truths of reason' on the other. 'Taken collectively,' Quine writes, 'science has its double dependence upon language and experience; but this duality is not significantly traceable into the statements of science taken one by one.'[15] And again, pushing the argument through: '[t]hat there is such a distinction to be drawn at all is an unempirical dogma of empiricism, a metaphysical article of faith.'[16] Rather, what we feel ourselves bound to accept as logical 'laws of thought' should instead be treated as particularly well-entrenched precepts, principles, or procedural guidelines which have so far proved to have great utility in working some manageable structure into the otherwise chaotic flux of incoming sensory stimuli. Any conflict at the edges of the fabric (that is, any 'recalcitrant' experience or item of anomalous observation-data) can always be assuaged by making certain adjustments elsewhere in the system, or by redistributing predicates and truth-values so as to restore equilibrium. At times this process may go so far – under pressure of conflicting empirical evidence – that it forces a change at or very near the logical heart of the fabric, that is to say, a revision to the hitherto sacrosanct 'laws of thought.' Thus '[a]ny statement can be held true, come what may, if we make drastic enough adjustments elsewhere in the system.'[17] And the same applies to those physical objects which we tend to think of as existing 'out there' beyond the boundary of the fabric, but which in fact – Quine argues – should rather be viewed as so many 'posits' whose being is dependent on our various theories, ontologies, descriptive languages, conceptual schemes, and so forth.

No doubt, instrumentally speaking, 'the conceptual scheme of science' is a useful tool 'for predicting future experience in the light of past experience.' But this should not deceive us into thinking that the objects posited by science – from quarks and gluons, via electrons, atoms and molecules, to the whole range of middling and large-scale objects – are more or less 'real' according to some *ultimate* (non-scheme-dependent) standard of objective truth. For what, after all, could such a standard consist in if not our current best notions of 'reality' derived from a mixture of commonsense wisdom plus other, more specialized (and preferably up-to-date) items of scientific knowledge? In which case – given the presumed non-finality or the always revisable status of science as we have it – there is simply no distinguishing between various 'posits' (from centaurs and Homer's gods, via numbers, mathematical classes or sets, to brick houses on Elm Street) in point of ultimate reality or truth. Rather, as Quine puts it:

> [p]hysical objects are conceptually imported into the situation as convenient intermediaries – not by definition in terms of experience, but simply as irreducible posits comparable, epistemologically, to the gods of Homer. For my part I do, qua lay physicist, believe in physical objects and not in Homer's gods; and I consider it a scientific error to believe otherwise. But in point of epistemological footing the physical objects differ only in degree and not in kind. Both sorts of entities enter our conception only as cultural posits.[18]

The same applies to those relatively abstract entities – 'classes and classes of classes and so on up' – which constitute the object-domain of mathematical reasoning, and whose status has long been a topic of dispute among Platonists, conventionalists, intuitionists and others. For Quine, such disputes must be seen as altogether pointless, presuming as they do that there exists some answer – some ultimate, non-scheme-dependent answer – to questions concerning the 'reality' (or otherwise) of numbers, sets, classes and so on. 'Epistemologically,' he writes, 'these are myths on the same footing as physical objects and gods, neither better nor worse except for differences in the degree to which they expedite our dealings with sense experience.'[19] In which case we had better just acknowledge that there is nothing – at any rate nothing 'in the nature of things,' whether physical objects, mathematical classes, centaurs, or Homer's gods – that could possibly justify our ranking those objects on a scale (or on different scales) according to their various well-defined orders of real, ideal, hypothetical, fictive or presumptively mythic existence. Quite simply, they are all 'cultural posits' which 'enter our conception' just in so far as they allow us to construct some plausible framework for relating 'experience' (or sensory stimuli) to the various other theories, posits, and items of belief that make up our current ontological scheme.

Now the really odd thing about all this is that Quine can be a hard-headed physicalist as concerns those sensory 'stimuli' while none the less adopting a wholesale contextualist approach with regard to what counts, on some given interpretive scheme, as a candidate real-world object or entity. Indeed, this might be called the third dogma of empiricism, one that supposedly makes short work of the old analytic/synthetic or scheme/content dualism, but which opens up a yet more problematic gulf between this or that item of stimulus–response and the overall 'fabric' or system of beliefs wherein such items have to be somehow accommodated. It is here – in the space between causal stimulus and cultural 'posit' – that a wide range of current anti-realist or cultural-relativist doctrines manage to insert their claim, often in terms that avowedly disown any such characterization. Thus, for instance, Richard Rorty can declare himself a full-fledged 'realist' in the sense that he accepts (1) the existence of a real-world or mind-independent object domain, (2) the fact that this world impinges directly on our nerve-ends or stimulus–response mechanisms, and (3) the necessity of defining 'truth' with reference to just this causal link between sensory stimuli and contents of belief. But he can then go on to claim that items 1–3 above have absolutely no bearing – or exert no ultimate constraint – upon the range of interpretive choices that exist with respect to our various theories, descriptions, world-hypotheses and so on. For on this view it is a species of category-mistake to suppose that whatever transpires in the causal interaction between world and mind is in any way decisive as regards the construal to to be placed on those various items of stimulus-input. That is to say, there is no passage from a Quinean physicalist (or naturalized) epistemology to an account of how certain stimuli are linked to certain perceptions, how these latter play a role in certain (more or less reliable) observation-sentences, and so on up to the stage of rational belief- or theory-formation.[20]

Thus, as Rorty sees it, there are two different kinds of 'aboutness' relation involved in the ascription of meanings and beliefs to people who can be safely be assumed to interact with their immediate physical environment in the way that Quine describes. One is the externalist viewpoint adopted by a Quinean 'radical interpreter' who *ex hypothese* lacks any inside knowledge of what informants may think or believe, and who therefore has to work on the basis of observing just such causal interactions, along with their responses (whether active or verbal) to various incoming stimuli. Then there is the kind of process that we ourselves and (presumably) other people engage in when trying to interpret, describe, explain or rationally justify items of belief. Thus the first is a basically determinist theory having to do with 'an aboutness relation that ties *s* [the relevant item] to its objects,' or which supposes a direct causal link between real-world occurrent states of affairs and beliefs (or belief-dispositions) on the part of real-world situated agents. The second is a quite different sort of theory, one that is essentially normative in character and has to do with 'the inferential relations between our belief that *s* and our other beliefs.'[21] Now according to Rorty it is simply an error – and the source of much epistemological confusion – to conflate or assimilate these two kinds of 'aboutness' relation. For the causal theory is just the sort of thing that applies when we are taking the externalist view of how people acquire or manifest various states of belief. It is therefore wholly devoid of normative implications, issuing as it does from a standpoint of studious non-commitment as regards the likely reasons, evidential criteria or justificatory grounds that those informants might be expected to produce by way of argumentative support. And conversely, there is nothing in the normative account – the 'view from inside' – that could claim good warrant through a causal appeal to beliefs reliably attained by exposure to the right kinds of physical stimuli. Such beliefs are entirely unconstrained by the aboutness-relation which, in Rorty's phrase, 'ties *s* to its objects.' Rather, they involve a rational (coherence-maximizing) method for ensuring that our various beliefs hang together in a broadly satisfactory way, and also for imputing beliefs to other people on the assumption that – more likely than not – they will be applying the same principle.

What this amounts to is a version of the double-aspect theory, familiar at least since Kant, which requires a careful separation of realms between issues of causal explanation on the one hand, and issues of rational justification on the other. More recently the theory has been given a new lease of life by Donald Davidson's idea of 'anomalous monism' as means of resolving the mind–body problem, or at any rate helping to defuse the quarrel between physicalist and mentalist arguments.[22] Such problems arise from a failure to perceive that these two conceptions need not get into conflict just so long as one accepts that they involve quite different kinds of understanding, the one aimed toward a causal-explanatory (perhaps a neurophysiological) account of mental processes and events, while the other is concerned to make rational sense of our own and other people's acts, intentions, meanings and beliefs. So, from this point of view, there is no difficulty about adopting even a strong (for example, central-state materialist) position on the issue of mind/brain identity while also holding that the other (intentional or belief-related) mode of talk is perfectly valid – indeed indispensable – when applied in the right sorts of context. Hence Davidson's theory of 'anomalous monism': *monist* in so far as it it rejects the idea of any radical dualism between mind and body, but *anomalous* in so far as it rejects any version of the hardline determinist argument according to which mental events are mere epiphenomenal illusions brought about by the working of physical laws whose better understanding would leave no room for such otiose 'folk-psychological' talk.[23] Thus we can all be as realist as we like in the external (causal-explanatory) mode and yet leave room for a strong interpretivist or hermeneutic approach when construing people's utterances, thoughts, and intentions on the basis of a Davidsonian 'principle of charity' which brings them out with as many true or rational beliefs as possible.[24]

In the same way, Rorty suggests, we can surely go along with the realist's case for the existence of an objective, mind-independent, causally efficacious 'external world' which is not just a construct – as the incautious relativist might have it – out of our various language-games, conceptual schemes, interpretive conventions or whatever. However, this gives no absolutely no grounds for holding that our own or other people's beliefs concerning that world are in any way constrained – or their truth-value decided – by the causal link that 'ties *s* to its objects.' Rather, such beliefs are arrived through an ongoing process of interpretive trade-offs or a Quinean pragmatic 'redistribution' of truth-values and predicates. Thus for Rorty, as likewise for Davidson and Quine, there is no sure route – perhaps no route at all – from a naturalized (causal) epistemology to a theory of rational belief-formation that would take due account of this process and the normative values it brings into play. These latter belong to the 'inside view,' that which we occupy in our role as self-conscious, reflective subjects for whom the word 'true' is a 'term of praise' applied to beliefs which optimize our sense of overall purpose and coherence. From the externalist viewpoint, conversely, 'true' is a term which properly applies to just those utterances or items of belief which display the right kind of causal history as tracked by a Quinean 'radical interpreter' with access to the relevant information sources but lacking any kind of privileged epistemic warrant. What is more, this restriction is equally in force when we adopt an externalist perspective on our own processes of knowledge-acquisition. For in this case also, as Rorty puts it, we should resist 'the urge to coalesce the justificatory and the causal story,' since they involve entirely different (incompatible) orders of truth-claim.[25] Epistemologically speaking, 'there is nothing more for us to know about our relation to reality than we already know' on the causal account, that is, as a result of simply observing how we and other people respond to certain stimuli under certain ambient conditions.[26] To jump from this to a normative 'story' about what we ought or ought not to believe as a matter of rational inference is just the kind of error that philosophers typically fall into when they fail to take the pragmatist's simple point about the various senses of 'truth.' Thus (Rorty again): 'the understanding you get of how the word "true" works by contemplating the possibility of a Tarskian truth-theory for your language is utterly irrelevant to the satisfaction you get by saying that you know more truths today than you did yesterday, or that truth is great, and will prevail.'[27] Least of all should we suppose that the causal sense of the term – applying as it does to the rock bottom level of unmediated stimulus–response – could possibly figure in a normative or justificatory account of our various holdings-true.

Now it might well be said that Rorty is untypical in pushing so far with the incompatibilist argument that it severs every link between the two kinds of 'aboutness'-relation, that which 'ties *s* to its objects' in a straightforward causal way and that which concerns 'the inferential relations between our belief that *s* and our other beliefs.'[28] Clearly one motive for adopting this line is Rorty's desire to avoid being characterized as a downright anti-realist (or cultural relativist) while still leaving maximum scope for his claim that we can 'redescribe' the world in as many different ways as there exist different language-games, 'final vocabularies,' metaphors we can live by, and so forth. Nevertheless Rorty's route to this wished-for *denouement* is one that follows a path well beaten by other analytic (or 'post-analytic') thinkers, Quine and Davidson among them. In Quine it takes the form of a naturalized or physicalist epistemology, along with a range of inter-connected theses – semantic holism, meaning-variance, ontological relativity, the theory-laden character of observation-statements, the underdetermination of theory by evidence – all of which conduce to a radical split between the baseline (causal) and higher-level (normative or interpretive) stages of belief-acquisition.[29] In Davidson, likewise, the argument works out as a doctrine of direct realism which allows us 'to reestablish unmediated touch with the familiar objects whose antics make our sentences and opinions true or false,' joined to an all-purpose Principle of Charity according to which we have no choice but to interpret other people's sayings and beliefs on the assumption that they must be counted 'right in most matters,' and hence sufficiently in agreement with us for those sayings and beliefs to be rationally construable.[30] So it is that Rorty can enlist Quine and Davidson in support of his case for combining an outlook of sturdy commonsense realism *vis-à-vis* beliefs and objects-of-belief with the idea of their always being open to creative 'redescription' in line with our current best interests or most favoured cultural values.

From here it is a short step to those varieties of jointly 'post-analytic' and depth-hermeneutical thinking which look to Heidegger for a way beyond the apparent dead-end – or the seeming triviality – of truth-definitions of the type proposed by thinkers in the mainstream analytical tradition.[31] For these latter reduce either to a Tarskian notion of truth as a purely definitional construct ('"snow is white" is a true statement if and only if snow is white') or to a crudely empiricist conception according to which the truth-content of sentences, statements, theories and so on should best be construed as a matter of 'surface irritations' – sensory stimuli of different kinds – plus whatever is needed in the way of Quinean across-the-board pragmatic adjustment. So it is not hard to see why some recent commentators have thought this a somewhat impoverished account of what goes on when we actually engage in the process of trying to figure out meanings and beliefs. One response is to urge that this whole way of thinking is hopelessly in thrall to an outworn epistemological paradigm which conceives 'truth' in terms of correspondence (of accuracy, factual content, propositional form, or whatever), and which thus remains closed to that other, more authentic or primordial dimension where truth is a matter of unconcealment, of the bringing-to-light of a wisdom concealed by the accretions of so-called 'Western metaphysics.'[32] On this view, philosophy has much to gain by the turn toward Heidegger and depth-hermeneutics, even if tempered – as in Rorty's case – by a measure of pragmatist skepticism as regards Heidegger's more portentous or *echt*-ontological pronouncements. To another way of thinking, no such drastic remedies are called for, but rather a careful reengagement with issues of truth, meaning and interpretation that have so far not been adequately treated by philosophers in the broadly analytic camp. On this view, conversely, we had much better stick with the resources of conceptual analysis and critique, whatever the problems hitherto encountered in applying those resources to particular questions of epistemology and interpretation-theory.

Thus, for instance, Simon Evnine remarks of Davidson that he seems to have 'two quite different projects …, [o]n the one hand a causal, explanatory project … [o]n the other an interpretative, hermeneutic project,' and that he fails to bring them together in any convincing or coherent way.[33] This problem itself has two aspects, the one concerning Davidson's idea of 'anomalous monism' (which in effect leaves it entirely unexplained just how we are to conceive the mind–body issue or the question of intentional *versus* physicalist accounts of human action), the other concerning the Davidsonian Principle of Charity and just how it relates to his otherwise thoroughly causal or determinist theory of belief-acquisition. For clearly there is a sense in which people (ourselves and others) *can* indeed be wrong about a whole range of things – whether motives for action, grounds for scientific belief, items of everyday-commonsense 'knowledge,' interpretive ascriptions of meaning or belief, sensory perceptions, inferential reasoning on this or item of evidence, etc. – where the Principle of Charity would bring them out right merely as a matter of *a priori* warrant for the claim to understand them on the basis of a shared rational or truth-optimizing strategy. In such cases what is needed, rather, is a jointly causal and normative epistemology which explains (1) how such errors come about, (2) why we are justified in viewing them as errors, and (3) what should count as an adequate reason for proposing some alternative – better informed or scientifically more adequate – approach to the issue.[34] Otherwise Charity will tend to work out as a kind of all-purpose hermeneutic pretext for counting people right (or rationally justified) no matter what they believe just in order to ensure that we don't lose touch and find ourselves forced into notions of radical paradigm-incommensurability or meaning-variance across different languages, cultures, 'conceptual schemes' and so forth.

This is, of course, where Davidson takes issue with Quine, Kuhn, Feyerabend, Whorf and other proponents of a relativist doctrine with just that unfortunate consequence.[35] But it is not at all clear that Davidsonian Charity is much better placed in this respect. For if consistently applied, then it will always come out *either* by counting other people 'right' on their own terms as a matter of general (truth-optimizing) principle, *or* by simply reconstruing their utterances on the 'charitable' basis that they must agree with us in most of what they say and believe, and hence that we can safely (justifiably) impute meanings and beliefs in large-scale accordance with our own ideas of rational, truth-preserving warrant. In which case, again, the way is wide open to a strong-redescriptivist (Rorty-style) argument that interpretation in some sense goes 'all the way down,' since there is nothing in the bottom-line causal theory of how beliefs hook up with objects-of-belief that could possibly serve to fix or individuate higher-level items of belief-content, propositional attitude, ontological commitment and so on. Thus the hermeneutic 'turn' in some recent readings of Davidson is one that takes off from precisely what Evnine remarks (in the passage cited above) as the split between his causal-explanatory project and a charity-based theory of belief attribution whose normative or justificatory criteria are completely uncoupled from such appeal to our causal transactions with the world. One very welcome consequence – so Rorty argues – is to cut down the role of epistemology (or theory of knowledge) to a point where it is finally squeezed out between the claims of a naturalized causal etiology of beliefs and a hermeneutic theory premised on the notions of meaning-holism, Davidsonian Charity and the pragmatist's idea of 'truth' as whatever merits that name by our present cultural-interpretive lights.[36] Of course, Rorty has no problems in reconciling this with Tarski's idea – likewise interpreted via Davidson – that a theory of truth for any given language should amount to just a formal recursive device which can be seen as simply canceling through for all practical purposes, and which therefore (arguably) lacks any kind of substantive ontological or epistemological content. So, on this account, the main drift of recent post-analytic philosophy has been to drive home the hermeneutic message: that epistemology is a hopelessly failed endeavor, since 'knowledge' and 'truth' amount to no more than our currently favoured ways of interpreting whatever is given at the basic level of unmediated (causally determined) stimulus–response. In which case clearly the sole remaining task for philosophy of mind and knowledge is to follow Rorty's lead and devote its main efforts to deconstructing all those grand theories erected by philosophers – from Descartes and Kant to the logical empiricists – who have laboured under the delusion that epistemology had other, more important or constructive work to do.[37]

3

As I have said, this message has not gone unchallenged among those who take a more sanguine view of the epistemological enterprise, but who reject the idea of a strict separation between externalist (causal-explanatory) theories on the one hand, and normative (reason-based or justificatory) arguments on the other. Thus, according to William Child, 'the idea that a normative, interpretative story is itself a causal story is essential for understanding the form of reason explanation.' Moreover, 'it is also essential for understanding the simple realist thought, that it is possible for us to have thoughts about, and knowledge of, "an objective public world which is not of our own making".'[38] This last phrase is quoted from Davidson and may be taken to mark the latter's dissent from any version of the strong-descriptivist (Rorty-style) argument that would find no room for an 'objective public world' except as a kind of noumenal posit, a reality that exists (by very definition) beyond whatever we can know or believe concerning it, and which otherwise enters out conceptions merely as a useful shared point of reference in the language-game of commonsense realism. All the same, it is far from clear that Davidsonian 'objectivity' amounts to anything more robust or secure than Rorty's pragmatist recasting of 'realism' as whatever is (currently and contingently) 'good in the way of belief.' For the effect of decoupling the causal and the normative conceptions of truth is to render the one entirely devoid of justificatory principles or grounds, while the latter floats free in a generalized willingness to interpret beliefs on a wholesale principle of charity which makes no distinction between those arrived at by reliable (knowledge-conducive) means and those adopted in line with existing assumptions or habits of thought. Thus perhaps it is the case, as Davidson writes, that a propositional (or belief-based) psychology 'cannot be, or be incorporated in, a closed science' of the type aspired to by physical theory and those branches of cognitive psychology that take their model from the physicalist paradigm of law-governed causal explanation.[39] But the problems are just as great if one swings right across to the opposite (hermeneutic or strong-descriptivist) extreme, or indeed if one adopts a position – like Davidson's 'anomalous monism' – which on principle refuses to perceive any possible conflict between beliefs brought about by causal interaction with the world and beliefs entertained as a matter of maximal coherence with our own or other people's convictions, world views, truth-conditional hypotheses and so on. For this gives Rorty just the argument he needs for declaring a complete severance between 'truth = whatever shows up in our unmediated causal interactions with the world' and 'truth = whatever we believe as a matter of cultural preference plus least resistance to incoming stimuli under this or that favoured (belief-preserving) interpretation.'

My point is that Davidson's doctrine of anomalous monism in effect gives us the worst of both worlds: a reductively physicalist account of causal explanation, and a theory of interpretive 'charity' which equates truth (or rationality) with a precept of across-the-board optimal adjustment between various ontological posits and items of belief. Thus, in Quine's words, '[e]ach man is given a scientific heritage plus a continuing barrage of sensory stimulation; and the considerations which guide him in warping his scientific heritage to fit his continuing sensory promptings are, where rational, pragmatic.'[40] What drops completely out of sight on this account is the process of 'rectification and critique' that plays such a crucial role in Bachelard's epistemology, and which involves a very different understanding of the relation between sensory 'promptings,' on the one hand, and the 'heritage' of received scientific thought on the other.[41] For Bachelard, this process cannot be reduced (as Quine would have it) to a matter of merely pragmatic adjustment, one that redistributes predicates and truth-values across the entire 'fabric' of existing beliefs so as to preserve maximum coherence with least disruption to cherished items. 'Conservatism figures in such choices,' Quine remarks, 'as does the quest for simplicity.'[42] But in that case it becomes hard to explain how science could ever have confronted a challenge to the orthodox wisdom of its time, or why 'normal' science (in Kuhn's terminology) should ever have found itself thrust into a phase of 'revolutionary' turmoil.[43] For there would always exist a whole range of alternative possibilities for reconstruing any item of 'recalcitrant' evidence so that it agreed with the body of accepted scientific lore. Or again, one could always tweak a few threads here and there in the overall fabric so as to accommodate the rogue item while avoiding any challenge to deeply entrenched beliefs.

Such would be the 'rational' line of least resistance as Quine describes it in 'Two Dogmas of Empiricism,' that is, the most pragmatically acceptable way of coping with anomalous data. However this argument fails to account for those decisive 'epistemological breaks' which, according to Bachelard, mark the transition from metaphor to concept, or from the stage of prescientific image and 'reverie' to the stage of adequately theorized scientific knowledge. Above all it evades the central issue of just why scientists should at times have broken so far with consensual norms as to postulate theories – such as relativity or quantum mechanics – which at first seemed completely at odds with all standards of rational scientific warrant. Now one might think that Quine is well placed to explain such episodes since it follows from his twin theses of ontological relativity and radical meaning-variance that *everything* is in principle always open to revision, from the construal placed upon incoming sensory stimuli to those putative logical 'laws of thought' that occupy the centre of the fabric. Indeed, quantum mechanics is Quine's chief example of the way that new developments in the physical sciences (for example, the wave/particle dualism) may force changes in the ground-rules themselves, as, for instance, by requiring a suspension of excluded middle or bivalent truth-values. Elsewhere, oddly enough, Quine retracts this claim by arguing that it could *never* be the best (most rational) option to relinquish those ground-rules in response to some set of anomalous or conflicting data.[44] For then – as Aristotle pointed out – we should find ourselves in the unfortunate predicament of endorsing both of two contradictory statements, and thus (by implication) asserting every statement that could ever be made on any conceivable topic.

No doubt there are deep problems with the interpretation of quantum mechanics, problems that have led some physicists, philosophers and formal logicians to propose an alternative (that is, non-bivalent or many-valued) logic in order to save scientific appearances.[45] However, there is also reason to think that this is a somewhat desperate stopgap remedy adopted for want of any better understanding of events in the quantum domain. My point, more generally, is that *none of these issues could ever have arisen* – from the break with 'classical' (pre-1900) ideas about the subatomic structure of matter to the further reaches of current debate – if those concerned had simply followed Quine's advice and finessed any conflicts by revising the import of observation-statements or redistributing truth-values wherever required. In that case normal science would continue pretty much undisturbed by the occasional (short-term) problem of adjustment between incoming sensory 'stimuli' and the background 'heritage' of scientific lore. What cannot be explained on Quine's account is the emergence of a theory – such as quantum mechanics – which not only transformed the fundamental concepts of physical science but created equally fundamental problems regarding its own interpretation, problems that are still (some hundred years on) just as far from being resolved.[46] It is therefore something of an irony that a good many currently *à la mode* doctrines in philosophy of science – ontological relativity, radical meaning-change, paradigm-incommensurability, anti-realism in its various forms – are assumed to find support from quantum mechanics, or at least to reflect the generalized uncertainty attaching to quantum phenomena. This is probably one reason for the widespread acceptance of Kuhn's ideas about 'revolutionary' science, not to mention the curious reversal of priorities by which some commentators (Rorty among them) have treated it as just the sort of thing that should always go on if we don't want to be stuck with boring old 'normal' science.[47]

However, there are several things wrong with the equation 'quantum mechanics = undecidability = truth as relative to whatever framework (system of measurement, paradigm, conceptual scheme, language-game, discourse and so on) one happens to adopt.' First, quantum mechanics assigns a precise value – that given by Heisenberg's constant – to the margin of uncertainty obtaining for any given measurement in any given case. Second, it is an error (and one with absurd consequences, as witness the strictly unthinkable predicament of Schrödinger's cat) to extrapolate directly from these paradoxes in the quantum domain to objects and events at the non-quantum (that is, macrophysical) level.[48] Third, whatever the problems encountered in making sense of such phenomena, they are problems that arise precisely from the effort to stretch our minds around a range of well-attested experimental findings – such as those of wave/particle dualism or non-local interaction – which are *not* just artifacts of some chosen paradigm, observation-language, theoretical framework and so on. And this despite the fact (also well-attested) that the so-called 'collapse of the wave-packet' into one or another state is in some sense produced, or its outcome decided, by the act of observation/measurement.[49] For there is no good reason – anti-realist prejudice aside – to derive the conclusion that quantum 'reality' is whatever we make of it according to this or that interpretive scheme. Rather, it is a matter of allowing for effects – like that of observation on the system observed – which would themselves have a place in any complete (so far unobtainable) description of the various physical factors involved. In short, these are – *pace* Bohr and his disciples – problems of a basically epistemological rather than an ontological nature.[50] That is to say, it is at least more rational to suppose that they reflect our current limited knowledge of events at the quantum level, and not some ultimate mystery at the heart of things that places such events forever and intrinsically beyond reach of scientific understanding. Thus what decides the truth of any given hypothesis – as between, say, the standard Copenhagen view and Bohm-type 'hidden-variables' theories – is whether or not it provides an accurate description of just those real-world events, irrespective of any beliefs that we or others happen to hold concerning them.[51]

For the anti-realist, conversely, it cannot make sense to posit the existence of 'verification-transcendent' truths, items of fact (of rational warrant or logical necessity) that would render our beliefs determinately true or false even in cases where we didn't know enough to ascertain the correct answer, or where currently available proof-procedures (for example, in mathematics) were inadequate to the task in hand, or where the evidence gave no means of settling the issue one way or another.[52] Thus for some analytic philosophers – notably Michael Dummett – 'warranted assertability,' not truth, is the most that we can justifiably hope for in assessing the content of various statements in mathematics, science, history and other disciplines of thought.[53] Such statements can only be evaluated in terms of our current best methods for bearing them out, whether through processes of mathematico-logical deduction, empirical evidence-testing in the physical sciences, or reference to reliable source-texts and archives in the case of historical research. What cannot be sustained – so this argument holds – is the notion of truth-values that possess an objective, real-world, evidence-transcendent or non-belief-relative status. Still less can the realist have grounds for asserting that we might just be wrong about a whole range of items – from mathematical theories to scientific or historical truth-claims – that come out true according to our best procedures of verification, but which are none the less false when measured against the way things stand 'in reality.' For this involves us in the strictly nonsensical claim (at least as the anti-realist sees it) that we might have grounds for both asserting some putative truth and at the same time asserting – with reference to some higher, more objective tribunal – that our assertion could always turn out false when subjected to more rigorous examination or when all the evidence was in. Quite simply, there is nothing that can warrant such truth-claims in the first place except their conforming to acceptable standards of logical procedure, disciplined enquiry, observational warrant, investigative scholarship and so forth. Any appeal beyond those standards – such as the realist's typical appeal to verification-transcendent truths – is one that *by very definition* could never be made good within the limits of knowledge or existing proof-procedures at any given time. In which case, the anti-realist argues, it is a mere a dogma of naive 'metaphysical' realism that we can well do without, since in giving it up we avoid all the skeptical problems that result from making this otiose distinction between 'truth in so far as we can possibly know it' and 'truth as determined by objective standards quite aside from our present best knowledge.' For it is just because realists insist on that distinction that they set themselves impossibly high standards for the verification of belief, and are hence – very often – driven into postures of reactive skeptical doubt.[54]

In Dummett's case, the route to this conclusion started out from intuitionism in philosophy of mathematics – the case that truth-values apply only to theorems for which we possess some adequate (decidable) proof-procedure – and then led on to kindred claims with respect to science, history and other fields of enquiry.[55] However, it is not hard to see why quantum mechanics – at least on the standard Copenhagen interpretation – has often been invoked in support of similar anti-realist positions. Thus it is argued (1) that quantum phenomena (such as the wave/particle dualism or the collapse of the wave-packet) cannot be assigned a determinate value apart from our acts of observation/measurement; (2) that this entails the impossibility of positing a 'real-world' quantum domain beyond such effects of observer-interference or outcomes contingent on the kind of experiment one chooses to conduct; and moreover (3), that it may also mean suspending certain axioms of classical logic (such as bivalence or excluded middle, if not perhaps non-contradiction) since these cannot apply to 'undecidable' instances like those mentioned in item 1 above. However, as I have said in connection with Quine, this whole line of argument is oddly out of touch with the actual development of quantum mechanics and the kinds of conceptual challenge it has posed during the past near-century of intensive discussion and research. For the quantum theory could never have emerged and continued to raise such acute interpretive problems if scientists had actually worked on the premise that nothing was in principle immune from revision, whether our conception of a real-world (mind-independent) physical domain or the most elementary ground-rules of logical thought. Had they taken this view – equating what is 'rational' with whatever is pragmatically best (or least vexatious) in the way of belief – then they would always have reinterpreted any recalcitrant data so as to minimize the extent of their conflict with existing scientific theories. Dummett has his own differences with Quine on these and related issues.[56] But if one puts together his anti-realist position and his case for the revisability of logic in response to certain (for example, mathematical or quantum-theoretical) dilemmas then it is no great distance to Quine's ideas of ontological relativity and radical meaning-variance. And from here the way is open to yet more extreme Rorty-type arguments for the 'reality'-constitutive powers of language, or Kuhnian/strong-sociological claims for the paradigm-dependent or culture-specific character of scientific theories.

4

I have argued that one alternative to this whole way of thinking is the approach to epistemology and philosophy of science developed by thinkers in the 'other' (Continental) tradition such as Husserl and Bachelard. What this approach has to offer, more specifically, is a way of treating these issues that combines analytical rigor and precision with a phenomenological account of the various emergent conflicts and resistances – 'obstacles' to thought, as Bachelard describes them – which have marked every major stage of advance in the modern physical sciences.[57] But in so doing it avoids the retreat to any kind of naive psychologism or any version of the 'subject-centered' appeal to Cartesian ideas of truth vouchsafed through privileged epistemic access. Rather, it involves a critique of precisely such ideas since they 'tend toward fulfilment in the realm of the image,' and hence block the way toward more adequate modes of scientific conceptualization.[58] This applies especially to those branches of modern science – such as quantum physics – whose conditions of emergence included the earlier break with intuitive or *a priori* notions of truth in fields like mathematics and non-Euclidean geometry. However, this process can never be treated as a simple *fait accompli*, a once-for-all passage beyond the naive idea that phenomenal intuitions (in Kant's famous phrase) must 'brought under' adequate concepts. On the one hand it involves a *constant and repeated* questioning of truths that appear self-evident not only from the standpoint of commonsense (non-scientific) belief but also in keeping with received ideas of scientific theory and method. On the other, it cannot be conceived as somehow transpiring in a realm of absolute ideal objectivity cut off from the historically emergent processes of thought that led up to some crucial paradigm-shift, or (in Bachelard's terms) some decisive 'epistemological break' with previous scientific beliefs. It is here – in a region largely unexplored by rational reconstructions in the analytic style – that philosophy of science can seek to comprehend the dynamics of scientific theory-change and the various motivating pressures and conflicts that brought such changes about.[59]

Imre Lakatos made a related point (albeit from a different methodological perspective) when he modified Kant's dictum to read: 'philosophy of science without history of science is empty,' while 'history of science without philosophy of science is blind.'[60] To which it might be added – with reference to Husserl and Bachelard – that these two endeavors must at some point be linked by a common concern with the genesis and structure of scientific theories, that is to say, their emergence through a critical-reflective process of thought which always strives toward 'absolute ideal objectivity' yet can never be entirely detached from its origins in certain historically situated acts of consciousness.[61] In conjoining the terms 'genesis and structure', I am alluding to Derrida's early essay on Husserl, along with other texts (such as his book-length introduction to Husserl's *The Origin of Geometry*) which are likewise centrally concerned with issues in epistemology and philosophy of science.[62] For those texts stand out – contrary to received 'analytic' opinion – as among the most rigorous and searching accounts of the difficulties faced by any attempt, like Husserl's, to reconcile these two ('genetic' and 'structural') approaches while avoiding the twin temptations of naive 'psychologism' on the one hand and empty formalism on the other. When read alongside Derrida's commentary on Bachelard in his essay 'White Mythology', they offer what amounts to a sustained engagement with just those problems that have preoccupied mainstream analytic philosophy of science since the high days of logical empiricism.[63]

I can best make the point by citing a passage from 'Genesis and Structure' which captures the gist of Derrida's argument concerning the strictly undecidable order of priority between these terms. It seems to me that the passage will bear close scrutiny not only by those who equate deconstruction with a Rorty-style 'textualist' approach, but also by those in the analytic camp who consider Husserl to have opened the way to a form of psychologism thinly disguised by his talk of ideality, consciousness-in-general, transcendental (as opposed to empirical) psychology, and the like:

> If Husserl gives up the psychological route when confronted by all the difficulties of accounting for a structure of ideal meaning on the basis of a factual genesis, he no less rejects the logicizing conclusion with which his critics wished to corner him. Whether in the then current Platonic or Kantian style, this logicism was preoccupied above all with the autonomy of logical ideality as concerns all consciousness in general, or all concrete and non-formal consciousness. Husserl, for his part, seeks to *maintain* simultaneously the normative autonomy of logical or mathematical ideality as concerns all factual consciousness, and its original dependence in relation to a subjectivity *in general*; *in general*, but *concretely*. Thus he had to navigate between the Scylla and Charybdis of logicizing structuralism and psychologistic geneticism (even in the subtle and pernicious form of the 'transcendental psychologism' attributed to Kant). He had to open up a new direction of philosophical attention and permit the discovery of a concrete, but nonempirical, intentionality, a 'transcendental experience' which would be 'constitutive,' that is, like all intentionality, simultaneously productive and revelatory, active and passive ... Husserl will attempt to prepare an access to this common radicality through the diverse 'reductions, which are presented initially as neutralizations of psychological genesis and even of every factual genesis in general. The first

> phase of phenomenology, in its style and its objects, is structuralist, because first and foremost it seeks to stay clear of psychologism and historicism. But it is not genetic description *in general* which is disqualified, but only the genetic description which borrows its schemas from naturalism and causalism, and depends upon a science of 'facts' and therefore on an empiricism; and therefore, concludes Husserl, depends upon a relativism incapable of insuring its own truth; therefore, on a skepticism. The transition to the phenomenological attitude is made necessary, thus, by the impotence or philosophical fragility of geneticism when the latter, by means of a positivism which does not understand itself, believes itself capable of enclosure by a 'science-of-facts,' whether this be a natural science or a science of the mind. The expression 'worldly genesis' covers the domain of these sciences.[64]

I have quoted Derrida's text at considerable length since it brings together all the main lines of argument developed in the course of this essay. In brief, they have to do with (1) the relation between deconstruction and phenomenology, (2) that between approaches to philosophy of science in the two ('analytic' and 'Continental') traditions, and (3) the problems that result when analytic philosophers attempt to exclude all reference to the activity of thought – the conscious, reflective, or critical activity – concomitant with advances in scientific knowledge or episodes of scientific theory-change. This is what Derrida calls (following Husserl) a 'positivism which does not understand itself,' one that 'borrows its schemas from naturalism and causalism,' thereby becoming a 'science of facts' devoid of epistemological content or warrant. So it is that empiricism opens the way to those varieties of skeptical-relativist doctrine that typically arise when philosophy encounters the absolute limit (the inbuilt resistance to thought) of naive phenomenalist or sense-data theories. After all, as Derrida remarks elsewhere, 'empiricism has always been determined by philosophy, from Plato to Husserl, as *nonphilosophy*, as the philosophical pretention to nonphilosophy, the inability to justify oneself.'[65]

It is not hard to see how this remark might apply to epistemology and philosophy of science in the wake of logical positivism. For when that programme collapsed under the pressure of various skeptical critiques as well as through its own internal contradictions – and when Quine followed up with his famous attack on the two 'last dogmas' of logical empiricism – then the way appeared open to a range of relativist arguments which exploited the sheer inability of any such doctrine to explain or to justify our knowledge of the growth of scientific knowledge. Some philosophers (Feyerabend chief among them) saw this as a happy deliverance, a liberating break with all those old oppressive notions of scientific truth, method, rational procedure, inference to the best explanation, and so forth.[66] Others again – such as Quine, Kuhn,

Davidson, Rorty (at times), and the later Putnam – went various ways around in attempting to formulate an approach that would hold the line against out-and-out relativism while retaining a commitment to what Derrida calls the 'schemas of naturalism and causalism.'[67] That is, they all took it that the best (indeed the only) way forward for epistemology and philosophy of science was one that treated knowledge as a direct product of our causal interaction with the world, with those objects and events whose impingement on our nerve-ends and the rest of our sensory-cognitive apparatus was enough to keep us reliably informed.

Hence Davidson's well-known advice on how to throw out the third (that is, the residual Quinean) dogma of empiricism and thus block the way to those skeptical worries that resulted from misplaced talk of 'paradigms,' 'frameworks,' 'conceptual schemes' and the like. 'In giving up the dualism of scheme and world,' he writes, 'we do not give up the world, but establish unmediated touch with the familiar objects whose antics make our sentences and opinions true or false.'[68] One can find many passages to similar effect in Putnam, Rorty, and indeed (ironically) in Quine himself, all recommending the adoption of a thoroughly naturalized epistemology which is meant to put us directly in touch with reality by cutting out the needless epistemological detour through concepts, categories, schemas, *a priori* intuitions, forms of representation, and suchlike. But this whole line of argument is fatally weakened by its reliance on a crudely empiricist idea of sense-data (or phenomenal experience) as somehow immediately given to the mind through a process of passive registration where nothing is required save a direct response to incoming sensory stimuli. Thus one need not be a card-carrying Hegelian – or even a disciple of Husserl – to note how implausible is Davidson's claim that the truth or falsehood of our various sentences (along with their role in our belief-systems, commonsense ontologies, scientific theories and so on) can be just a matter of 'unmediated' contact between our nerve-ends and whatever impinges on them from one moment to the next.

With Quine this leads straight on to the doctrine of ontological relativity since there is nothing in the nature of those bare sensory 'stimuli' that could possibly impose any limit on the range of stimulus-compatible schemes, frameworks, or theories. Logical considerations might narrow it down to some extent, for example, by excluding conflictual hypotheses or theories entailing some clear violation of bivalence or exluded middle. Yet of course Quine also makes the case, in 'Two Dogmas of Empiricism,' that these putative ground-rules of logic or 'truths of reason' are themselves always open to pragmatic adjustment under pressure from conflicting evidence, thus once again opening the way to full-scale ontological relativity. In Rorty, as we have seen, the causal theory of belief-acquisition goes along with a yet more extreme variety of cultural-relativist thinking, that is, the notion that sensory 'data' do *nothing whatsoever* to fix or to guide the various interpretations that may be placed on them according to this or that theory, language-game, or currently acceptable (scientific or other) communal form of life. And in late Putnam, the idea of 'internal realism' – that is, internal to this or that frame of reference – seems scarcely more promising: a Kantianism minus the transcendental subject (but still giving rise to all the problems of scheme/content dualism pointed out by Davidson), and a relativism that dare not speak its name but which lies open to the standard counter-arguments rehearsed by philosophers from Plato down.[69]

My point about these various dilemmas in the wake of logical empiricism is that they all take rise from what Derrida discerns – in the long passage on Husserl cited above – as the *inevitable* upshot of a positivist 'science of facts' that rejects any form of phenomenological or epistemo-critical enquiry. When he remarks that this approach 'borrows its schemas from naturalism and causalism,' the comment applies with particular force to attempts – such as those of Quine, Davidson, Putnam and Rorty – to overcome the problems of logical empiricism by simply cutting out all that otiose talk of 'truth,' 'correspondence,' conceptual adequacy, and so on, and adopting a straightforward physicalist theory of 'unmediated' stimulus–response. However, this solves none of the problems, as can be seen most clearly from their various periodic changes of mind (or, in Rorty's case, his nimble balancing act) concerning the realist/anti-realist debate and the associated issue of framework-relativism.[70] What leaves them in a quandary on both counts – and thus gives Rorty his tactical room for maneuver – is the narrowing of epistemological sights to a point where only sense-data (or 'stimuli') possess direct veridical warrant, and where everything else (from observation-statements all the way up to high-level theories) must therefore be regarded as framework-relative products of interpretation.

This upshot is manifest in Quine's 'Two Dogmas' and also – whatever his claim to have transcended such old 'metaphysical' habits of thought – in Putnam's pragmatist recasting of Kantian themes under the name of 'internal realism.'[71] Where it emerges most poignantly is in certain passages of Kuhn's *The Structure of Scientific Revolutions*, most of them occurring in the 1969 second edition postscript, but some of them to be found in the original (1961) text. What they reveal is Kuhn's distinct unease with the Quinean (physicalist) theory of belief-acquisition, and also, closely related to that, his sense that there might be something quite wrong with his (that is, Kuhn's) line of argument concerning such topics as paradigm-change, radical meaning-variance, the theory-laden character of observation-statements, the underdetermination of theories by evidence, and the idea that scientists somehow quite literally 'live in different worlds' from one paradigm to another. Thus, for instance:

> Is sensory experience fixed and neutral? Are theories simply man-made interpretations of given data? The epistemological viewpoint that has most often guided Western philosophy for three centuries dictates an immediate and unequivocal, Yes! In the absence of a developed alternative, I find it impossible to relinquish entirely that viewpoint. Yet it no longer functions effectively, and the attempts to make it do so through the introduction of a neutral language of observation now seem to me hopeless.[72]

Compare this with the passage from Derrida on Husserl cited above, and you will gain some sense of the extraordinary narrowness of historical and intellectual vision that led Kuhn to his doleful conclusion *vis-à-vis* the prospects for philosophy of science at the time. That vision is confined to just one major line of descent (broadly speaking, from Hume to his modern 'analytic' progeny) and, moreover, to just the kinds of problem thrown up by movements in the wake of logical positivism. Thus Kuhn's way of stating the issue – as a matter of imposing 'man-made' theories or interpretations on 'fixed and neutral' sense-data – is one that would appear remarkably crude to anyone possessing a knowledge of developments in the other main tradition of philosophic thought running from Kant to Husserl. They would also most likely want to challenge his claim that this has been the 'epistemological viewpoint' that has guided Western thought on such topics for the past three centuries, and which leaves no room for any 'developed alternative' despite the hopelessness (as Kuhn sees it) of all attempts to establish a neutral observation language. For there do exist other, highly developed alternatives – some of which I have discussed in the course of this essay – that avoid the predicament so plaintively evoked in the above passage from Kuhn. That is, they follow Husserl in rejecting the positivist idea of knowledge as a product of raw 'sense-data' which are then somehow linked up with concepts (hypotheses, theories, covering-law statements) through a process of so-called rational reconstruction that bears no relation to the processes of thought by which scientific discoveries actually come about.

This approach was carried on by the logical empiricists who introduced a whole range of further distinctions – for example, that between first-order (natural) and formalized languages, enshrined in the Carnapian doctrine of semantic ascent – with a view to expounding the structure and logic of scientific theories quite apart from the messy contingencies of scientific practice.[73]

But when that doctrine collapsed under skeptical pressure (especially from Quine's 'Two Dogmas') the way appeared open, via meaning-holism, to various proposals – Kuhn's among them – for relativizing truth to this or that framework of in-place consensus beliefs. Hence his two rhetorical questions ('Is sensory experience fixed and neutral? Are theories simply man-made interpretations of given data?'), questions that Kuhn clearly regards as inviting a negative response, but which he still feels driven to ask for want of any viable alternative approach. However, they will seem less pressing – or at any rate not so far beyond hope of an answer – if one looks outside the restrictive purview of debates within the analytic tradition, at least as standardly defined. For it will then become apparent that there exists a large body of work, some of it inspired by Husserl or Bachelard, that has engaged these issues at a high level of conceptual rigor and analytical precision while avoiding the various drastic dichotomies bequeathed by logical empiricism. Chief among these is the dichotomy imposed by its account of what constitutes a basic observation-sentence (grounded in empirical 'sense-data') and a properly formulated scientific theory (that is, one constructed out of just such data through standardized logico-deductive procedures of reasoning on the evidence). It is just this sharply disjunctive view of the relation between evidence and theory – along with the consequent reactive slide into various forms of positivism, naturalism, skepticism and relativism – that Husserl diagnosed as a 'crisis' besetting both the physical sciences and the very project of enlightened critical thought.[74]

Thus the great task for Husserl, in Derrida's words, was 'to open up a new direction … [to] permit the discovery of a concrete, but nonempirical, intentionality, a "transcendental experience" which would be "constitutive", that is, like all intentionality, simultaneously productive and revelatory, active and passive.'[75] By so doing it should yet be possible to reverse the tide of rampant unreflective positivism in the natural sciences and also the backwash that, according to Husserl, had produced the contemporary flight to a range of post-Nietzschean irrationalist philosophies. This task was by its very nature interminable, Derrida thinks, since it involved certain deep-laid antinomies (structure/genesis, concept/intuition, 'active' and 'productive' *versus* 'passive' or 'revelatory' modes of understanding) which would always at the limit defeat any attempt to reconcile their various conflicting claims. Hence also the vital distinction (as Husserl described it) between *exactitude* and *rigor* as aims characteristic of different sciences at different stages or depths of phenomenological enquiry. Thus:

> an eidetic descriptive science, such as phenomenology, may be rigorous, but it is necessarily inexact – I would rather say 'anexact' – due to no failure on its part. Exactitude is always a product derived from an operation of 'idealization' and 'transition to the limit' which can only concern an abstract moment, an *abstract* eidetic element (spatiality, for example) of a thing materially determined as an objective body, setting aside, precisely, the other eidetic elements of a body in general. This is why geometry is a 'material' and 'abstract' science. It follows that a 'geometry of experience,' a 'mathematics of phenomena' is impossible: this an 'attempt doomed to miscarry.' This means in particular, for what concerns us here, that the essences of consciousness, and therefore the essences of 'phenomena' in general, cannot belong to a structure or 'multiplicity' of the mathematical type. Now what is it that characterizes such a multiplicity for Husserl, and at this time? In a word, the possibility of *closure* .... What Husserl seeks to underline by means of this comparison between an exact and a morphological science, and what we must retain here, is the principled, essential, and structural impossibility of closing a structural phenomenology.[76]

So there is no question of finally resolving these issues in the philosophy of mathematics, geometry and the physical sciences, a point that Derrida himself underlines by reference to Gödel's incompleteness theorem and other related developments concerning the limits of computable or axiomatic proof-procedures. But there is also no question of relapsing into the kind of naive positivist attitude that would treat such issues either as belonging to a realm of abstract mathematical concern outside its own empirical remit or – in so far as they involve talk of Husserlian 'essences,' structures of consciousness, 'eidetic descriptions' and so on – as merely a species of misguided psychologism. For this doctrine goes so far in its rejection of 'metaphysical' talk that it effectively debars any genuine engagement with the processes of thought that accomplished those breaks with natural (commonsense-intuitive) understanding or received scientific ideas.

5

Even when it first appeared during the 1920s, Logical Positivism was philosophically regressive and oddly out of touch with recent scientific developments. These included the emergence of quantum mechanics with its sizable challenge – on whatever construal – to the positivist conception of knowledge as acquired through a passive registration of sensory data that could then be cast in propositional form and brought under some suitable set of generalized covering-law statements.[77] Small wonder that Kuhn was at a loss as to how this theory – in its later, logical-empiricist guise – could possibly provide an adequate basis for explaining the structure of scientific revolutions. Indeed, his whole approach to that topic is perhaps best viewed as the result of Kuhn's disillusionment with the programme of logical empricism, along with his failure to conceive or envisage any viable alternative account. Thus the 'structure' that Kuhn rather vaguely refers to is one that seems to exist in an abstract realm quite apart from the particular features – the tensions, resistances, conflicts of allegiance, struggles to reconcile theory and observation – which have marked periods of looming transition or pre-revolutionary ferment. It is not so much that Kuhn ignores these factors (in a sense, his book is about little else), but rather that he treats them from a viewpoint that excludes all concern with the processes of thought – at whatever conscious, unconcious, or pre-conscious level – through which they emerged as a obstacle or challenge to existing scientific beliefs.

In this respect, ironically enough, Kuhn's approach gives rise to something very like the old logical-empiricist split between history and philosophy of science, or what actually went on in the 'context of discovery' (including the false starts, failed experiments, anomalous data, ambiguous observations, dubious conjectures, and other such messy details) and what should go on in the 'context of justification' when philosophers seek to rationally reconstruct the logic of scientific enquiry.[78] Of course, Kuhn rejects this distinction in principle since he sees it as upholding the notion of science as a steady progress from stage to stage in the advancement of scientific knowledge. That notion has to go, he thinks, when we acknowledge the incommensurability of paradigms and the amount of semantic jerrymandering that has to occur before we can presume to rank different theories on a common evaluative scale, or claim to know more (compared with our scientific forebears) about historically shifting concepts such as 'mass,' 'gravity,' 'motion,' 'space,' 'time' and so forth. All the same he still adopts an externalist approach that treats these issues, so to speak, from outside and above. That is, he assumes that they can only make sense from the standpoint of a present-day narrative account which relates various episodes in the history of science without judging their truth-content and without the kind of detailed enquiring-back into their epistemological conditions of emergence that would offer grounds for contrasting them in point of descriptive accuracy, explanatory power, theoretical refinement and so on. In other words, Kuhn rejects one kind of meta-historical perspective – that which treats science as an onward and upward march toward standards of truth prescriptively defined by its own criteria of logico-deductive validity – only to espouse another such perspective according to which past and present theories must always be treated as relative to (or constructed by) this or that range of paradigm-specific beliefs. This approach may claim the liberal virtue – as with the strong sociologists – of *not* rewriting the entire history of science to date in accordance with our own present notions of truth, rationality, and method.[79] But in fact it operates on a kind of reverse-discrimination principle, discounting a whole range of past scientific achievements (and the often strenuous activity of thought which led up to them) as possessing no distinctive claim to validity or truth.

This results from Kuhn's highly generalized idea of what constitutes the 'structure' of scientific revolutions, that is to say, his habit of abstracting away from the particular intellectual contexts in which such revolutions occurred and the particular kinds of conceptual, theoretical, or epistemo-critical advance which brought them about. With the phrase 'epistemo-critical' I am once again referring to Bachelard's project of *rationalisme appliqué*, an outline of which I have offered already, and which offers – I would argue – a promising way forward from the dilemmas both of logical empiricism and of Kuhnian paradigm-relativism. More specifically, it shows how the process of scientific theory-change can be treated from a standpoint that takes due account of the thinking that produced such changes, but which none the less avoids any naive (philosophically and scientifically retrograde) appeal to *a priori* concepts, primordial intuitions, privileged epistemic access, or whatever. It is just this possibility that Derrida holds out in his reading of Husserl as constantly engaged in a strenuous attempt to navigate between the twin poles – the 'Scylla and Charybdis' – of 'a logicizing structuralism and a psychologistic geneticism.'[80] It is also what Derrida finds most revealing in Husserl's late text on 'The Origin of Geometry,' where he raises the question: how is it possible that an axiomatic-deductive science, one supposedly grounded in truths self-evident to reason, should yet have developed through various stages of historically datable discovery, from Euclid down to the emergence of alternative (non-Euclidean) geometries?[81] In the end Husserl cannot answer this question, caught up as he is – on Derrida's account – in a constant shuttling exchange of priorities between the rival claims of 'structure' and 'genesis.' On the one hand, 'it is always something like an *opening* which will frustrate the structuralist project. What I can never understand, in a structure, is that by means of which it is not closed.'[82] On the other hand, any too pronounced swing toward the opposite ('geneticist') pole will always run the risk of falling into a naive psychologism which fails to explain how sciences such as geometry, mathematics, or physics could ever have accomplished a decisive break with intuition-based or *a priori* conceptions of space, time and number.

In Husserl, this aporia takes the form of an alternating rhythm, a constant switching of priorities, such that these two projects of thought (roughly speaking, 'structuralism' and 'genetic phenomenology') coexist in a kind of tense but revealing mutual interrogative exchange. In Derrida, it is taken up and developed to a point where those tensions are yet more visible but where they are seen as a *strictly inevitable outcome* of critical enquiry into the structure and genesis of scientific truth-claims. Here again there is a pointed, by no means merely offhand or casual relevance when Derrida refers to the 'difficulties' posed for Husserl's philosophy by 'certain later developments of axiomatics and by Gödel's discoveries.'[83] What is in question is not some vaguely generalized notion of 'undecidability' that would take Gödel's theorem as a pretext for suspending all notions of objective truth, all criteria of right reason or standards of valid argument. Rather, it concerns the surprising fact that every formal system – for example, that of arithmetic – contains certain strictly undecidable propositions, axioms whose truth is a matter of stipulative warrant but which cannot be proven within the system. As Roger Penrose states it: 'no formal system of sound mathematical rules of proof can ever suffice, even in principle, to establish all the true propositions of ordinary arithmetic.'[84]

In this case it might be supposed that there exist any number of different arithmetics, each of them equally valid by its own formal criteria, since none of them can be *both* consistent *and* complete. However this is to mistake the import of Gödel's theorem, according to which mathematical truth – the truth of that theorem included – is 'something absolute, there to be discovered, rather than invented.'[85] In other words, the very fact that mathematical reasoning can establish the incompleteness of its own proof-procedures is itself evidence that we possess other sources of knowledge, sources that exceed any possible power of computation or demonstrative proof. What the theorem tells us, Penrose suggests:

> can be viewed in a much more positive light, namely that the insights that are available to human mathematicians – indeed, to anyone who can think logically with understanding and imagination – lie beyond anything that can be formalized as a set of rules. Rules can sometimes be a partial substitute for understanding, but they can never replace it entirely.[86]

In making this case, Penrose has his eye on the wider implications of Gödel's theorem, especially as applied to issues in the field of artificial intelligence, where it suggests – in line with the central thesis of his book – that no computer could ever be programmed to achieve such creative-intuitive insights. However, the main point here is that he shows very plainly why the theorem *cannot* be construed as somehow 'proving' that arithmetic is an arbitrary construct, that mathematical truths are purely conventional, that there exist any number of possible ways of performing some given calculation, continuing some sequence of natural numbers, and so forth. On the contrary, according to Penrose: 'it is within mathematics that we find the clearest evidence that there must actually be something in our conscious thought processes that eludes computation.'[87] For otherwise there could be no explaining how Gödel arrived at his (on the face of it) highly counter-intuitive result.

What the theorem thus demonstrates is the existence of truths – its own included – which are there to be discovered (not invented), but which lie beyond the reach of any rule-governed, algorithmic or computable process of thought. It is worth remarking, though Penrose doesn't make the point, that this issue about 'following a rule' – what counts as a rule? which of all the various conceivable rules that might apply in this or that case? how to place a limit on the regress from rules to rules-for-the-application-of-those-rules?, and so forth – is one that has been raised and most often pushed in a skeptical direction by various philosophers after Wittgenstein.[88] Indeed, it is one source of that drift toward forms of cultural-relativist thinking which has been such a prominent feature of recent 'continental' and 'post-analytic' philosophy alike.[89] So there is a further significance to Penrose's argument concerning the correct interpretation of Gödel's incompleteness theorem. What that argument shows, in brief, is that problems with the notion of 'following a rule' – whether genuine or spurious problems – are in any case strictly beside the point when it comes to instances of reasoning (such as Gödel's proof) which pass beyond the limits of computable or rule-governed calculative thought. Thus 'no such system of rules can ever be sufficient to prove even those propositions of arithmetic whose truth is accessible, in principle, to human intuition and insight – whence human intuition and insight cannot be reduced to any set of rules.'[90]

On the standard construal of this problem from Wittgenstein, the only resort is to a 'skeptical solution' (Kripke's phrase) which re-states the problem in a slightly sharpened form and then proposes that we set it aside by appealing to shared practices or communal procedures within which those rules just *are* the way we do things as a matter of received – for example, mathematical – convention. But of course this 'solution' solves precisely nothing if one asks (like Wittgenstein) what ultimate justification there could be for claiming that someone who said '2 + 2 = 5,' or who continued the number-series '2, 4, 6' with the sequence '11, 16, 29,' was in some fairly obvious way failing to grasp the arithmetical point. It seems to me that this is just a pseudo-problem produced by the standard Wittgensteinian move of rejecting all truth- or validity-claims except those sanctioned by the fact of their playing a role in some given 'language-game' or cultural 'form of life.' Nevertheless there is further support to be had from Gödel's theorem as Penrose interprets it, namely his argument that reasoning cannot be entirely rule-governed since we have access to a wide range of truths (in mathematics and elsewhere) which exceed the bounds of computability.

However, the chief point I wish to make is that Derrida's usage of 'undecidability' – in his reading of Husserl and elsewhere – is concordant with this understanding of the term and not with the kind of loose deployment to be found in some adepts of literary deconstruction, and likewise in the notion of 'postmodern' science entertained by thinkers such as Jean-François Lyotard.[91] On this interpretation it amounts to no more than a catch-all substitute for other ideas – 'uncertainty,' 'chaos,' 'paralogism,' quantum nonlocality, the wave/particle dualism or the limits of precise measurement, to take just a few from Lyotard – all of which supposedly signal the end of science as a rational, truth-seeking enterprise. For Derrida, conversely, the reference to Gödel occurs more than once at a specific juncture in his argument where the undecidability is that which results from a conflict between two *strictly indispensable but strictly incompatible* logics of enquiry.[92] In Husserl's case, this conflict has to do with the issue of priority between genesis and structure, an issue that Derrida is unable to resolve (any more than could Husserl), but which he presses to the point where it reveals just why there should exist this constitutive tension in the project of transcendental phenomenology. Moreover, it is a tension that will always emerge – so Derrida argues – when philosophy reflects on the conditions of possibility for scientific knowledge in general, and on the various stages of historical development that have marked its progress to date. For the logical empiricists these were to be treated as two quite separate disciplines, the latter concerned with history of science and the so-called 'context of discovery,' while the former pursued the more important task of reconstructing and testing scientific theories with benefit of rational hindsight, or with respect to their success or failure as measured by standards properly pertaining to the 'context of justification.' As I have said, this dichtotomy persists in the work of thinkers like Kuhn, those who reject the logical-empiricist programme but who still deploy the term 'structure' in a generalized sense that scarcely makes contact with the detailed processes of paradigm-change or the particular conflicts of theory and evidence that bring such changes about. The result is on the one hand to empty that term of all genuine descriptive-explanatory content, and on the other to leave an opening for just the kind of wholesale relativist approach (exemplified by the strong sociology of knowledge) that equates 'truth' with whatever holds good for some *de facto*, empirically determined community of belief.

6

Derrida's main concern in his work on Husserl is with the tension between structuralism and phenomenology, each being construed (as I have said) in a sense that goes beyond their normal application to recent European movements of thought. If 'a certain structuralism has always been philosophy's most spontaneous gesture,'[93] then equally it is the case that a certain phenomenology has always asserted its claim whenever thinking has turned toward questions regarding the history and genesis of the various sciences. In the context of Derrida's early (pre-1970) work, this issue was chiefly engaged by French phenomenology in the wake of Husserl and the emergent structuralist 'sciences of man' which defined themselves squarely against that previous, supposedly naive or 'humanist' approach to questions of truth, meaning and interpretation. These latter of course took their bearings from Saussurean structural linguistics, and proposed to extend that pilot science to a whole range of other disciplines, among them anthropology (Lévi-Strauss), narratology (Greimas, Todorov, early Barthes), poetics (Jakobson), psychoanalysis (Lacan), history of ideas (Foucault) and political theory (Althusser).[94] What they all had in common was a fierce repudiation of the earlier phenomenological paradigm which accorded priority to the human subject as a locus of thoughts, meanings, expressive intentions, or acts of will that by their very nature exceeded the grasp of any purely structural account. For some thinkers, mainly in the structuralist camp, this conflict of interpretations was such as to allow no possible compromise or via *media* between the two approaches. For others (Ricoeur and Merleau-Ponty among them), it required a patient and meticulous thinking-through of the various philosophical problems that arose in attempting to reconcile their different claims.[95] Where Derrida pursues a distinctive path is in showing that this tension between 'genesis' and 'structure' is one that *in principle* can never be resolved, but which none the less opens onto questions of the profoundest import for epistemology, philosophy of science, and the humanistic disciplines. That is to say, those questions will simply not emerge if one adopts *either* a purebred phenomenological approach which denies the constitutive role of pre-given 'structures' or concepts, *or* (conversely) a thoroughgoing structuralist approach that ignores the extent to which thought and language may exceed and transform those pre-existing structures or concepts.

Thus it is important to recognize – despite and against one common misreading of his work – that when Derrida 'deconstructs' the discourse of Husserlian phenomenology, he does so not in order to discredit its claims but in order to demonstrate the 'absolute and principled necessity' of carrying that project forward in accordance with its own self-critical precepts. The following passage makes this case with particular clarity and force. 'All these [i.e., Husserl's] formulations have been possible,' Derrida writes:

> thanks to the initial distinction between different and irreducible types of genesis and structure: worldly genesis and transcendental genesis, empirical structure, eidetic structure, and transcendental structure. To ask oneself the following historico-semantic question: 'What does the notion of genesis *in general*, on whose basis the Husserlian diffraction could come forth and be understood, mean, and what has it always meant? What does the notion of structure *in general*, on whose basis Husserl *operates* and operates distinctions between empirical, eidetic, and transcendental dimensions mean, and what has it always meant throughout its displacements? And what is the historico-semantic relationship between genesis and structure *in general*? is not only to ask a prior linguistic question. It is to ask the question about the unity of the historical ground on whose basis a transcendental reduction is possible and is motivated by itself. It is to ask the question about the unity of the world from which transcendental freedom releases itself, in order to make the origin of this unity appear.[96]

There is much that could be said by way of unpacking this densely argued passage, but I must here content myself with a brief commentary. First, it brings out the sheer distance that separates Derrida's approach to these issues from the 'structuralist' approach manifested in works such Foucault's *Les mots et les choses*.[97] I choose this particular text for comparison – and with a backward glance to what I said about Kuhn several paragraphs above – because it shows very clearly how 'a certain structuralism' affects the very project of historical enquiry into the various forms, stages and developments of scientific knowledge to date. That Foucault rejected the 'structuralist' tag as applied to his own work is I think more a case of strategic track-covering than a genuine or deep-laid difference of approach. For there are two chief features of *Les mots et les choses* (and other works of that period) which lend themselves to precisely that description. These are (1) his resolute assault on all notions of the subject, whether the Cartesian 'subject-presumed-to-know' or again, more specifically, the transcendental subject as theorized by philosophers from Kant to Husserl, and (2), his idea that all claims to knowledge or truth must be seen as products of a certain 'structural *a priori*,' one that has nothing whatever to do with any (pseudo-)transcendental mode of argument, but which conforms to the dominant order of discourse at any given time and hence undergoes a more or less drastic mutation with every change in that order.[98] No doubt Foucault pushes both arguments much further in a skeptical-relativist direction than would be acceptable to Kuhn or like-minded thinkers in the Anglo-American camp. What they do have in common is a notion of 'structure' that is so devoid of epistemological content – so far detached from the kinds of thought-process involved in particular instances of scientific theory-change – that there seems no choice but to give up seeking to explain such episodes and to treat them rather as periodic shifts in the prevailing paradigm or the dominant discursive 'order of things.' In which case there quickly ensue all the well-known problems of meaning-variance and paradigm-incommensurability, problems that Kuhn (to his credit) perceived more clearly than Foucault, but to which he could provide – as I have argued elsewhere – no more convincing an answer.[99]

It is in response to questions such as these that Derrida insists on the continuing necessity of thinking through the genesis/structure relationship as raised to a high point of critical consciousness in Husserl's writings. The various distinctions that he draws from Husserl in the above-cited passage ('worldly genesis' and 'transcendental genesis,' 'empirical structure,' 'eidetic structure,' 'transcendental structure', and so on) are unlikely to carry much weight with philosophers in the Anglo-American tradition. However, I would suggest, it is chiefly for want of such epistemo-critical resources that this tradition gave rise to a doctrine – logical empiricism – whose perceived shortcomings have opened the way to Quinean, Kuhnian and other varieties of skeptical-relativist thinking. That is to say, it engendered a series of well-known problems that were strictly insoluble on its own terms, among them problems concerning the relation between theory and evidence, the status of empirical or observation data, the validity of generalized (covering-law) statements, and the question as to what should count as scientific progress given the lack of any sure criterion for evaluating theories in point of observational accuracy or conceptual-explanatory power. So it was that Quine could propound his skeptical theses – ontological relativity, the underdetermination of theory by evidence, the theory-laden character of observation-statements – as a consequence of just these perceived failings in the logical-empiricist programme. From here it was but a short step to those varieties of full-fledged relativist doctrine (such as the 'strong program' in sociology of knowledge) that likewise trade on the radical divorce between logicist conceptions of scientific method devoted exclusively to the context of justification, and empirical enquiries into the history of science that expressly reject all normative criteria of truth, reason and progress. For the latter can perhaps best be seen as a reactive movement of thought, one that embraces the authority of the social (rather than the physical) sciences, and which does so precisely in a gesture of revolt against the narrowness – the lack of any genuine historical dimension – that typifies the dominant analytic approach in its rational-reconstructive mode.[100] However, the strong sociologists, like Kuhn before them, can be seen as reverting to just the kind of uncritical positivist outlook that comes of relying on the sheer self-evidence of sociohistorical or factual-documentary data. In short, their approach is still dictated by a version of the logical-empiricist dichotomy, one that rejects the logicist horn of this particular dilemma while scarcely questioning the methods applied and the results turned up by empirical research in the social sciences.

I should not wish to claim that these problems receive a definitive answer in the broadly 'Continental' (post-Kantian) tradition of thought that is represented on the one hand by Husserlian phenomenology, and on the other by that form of epistemo-critical *rationalisme appliqué* developed by thinkers such as Bachelard. Indeed, it is a chief point of Derrida's argument that these issues – for example, that of 'genesis' and 'structure' or the relation between metaphor and concept in the discourse of science and philosophy – cannot be resolved, but must always be addressed over again with each new stage in the history of thought. However this is *not* to embrace some kind of irrationalist doctrine, any more than Penrose adopts such a view when he takes it (on the evidence of Gödel's incompleteness-theorem) that there exist mathematical truths and powers of human insight 'beyond formal argument and beyond computable procedures.'[101] Derrida is coming at the same issue – albeit from a different philosophical angle – when he remarks that 'it is always something like an *opening* that will frustrate the structuralist project'; moreover, that 'what I can never understand, in a structure, is that by means of which it is not closed.' And again, citing some passages from Husserl's *Cartesian Meditations*:

> [t]he results of phenomenology are 'metaphysical, if it be true that ultimate cognitions of being should be called metaphysical. On the other hand, what we have here is anything but metaphysics, in the customary sense with which metaphysics, as "first philosophy", was instituted originally.' 'Phenomenology indeed excludes every naive metaphysics ... but does not exclude metaphysics as such.' For within the most universal *eidos* of mental historicity, the

> conversion of philosophy into phenomenology would be the final degree of differentiation (stage, that is *Stufe*, structural level or genetic stage).[102]

Clearly this passage takes aim against the use of 'metaphysics' (or 'metaphysical') as a routine term of abuse, one that has often been applied by analytic philosophers to signify the kind of meaningless talk that they think typical of work in the 'other' (post-Kantian Continental) tradition.[103] For it is precisely that undifferentiating usage – the failure to distinguish 'naive' metaphysics from 'metaphysics as such' – that characterized the discourse of logical positivism and also (in many cases) the thinking of its logical-empiricist heirs and successors. As a result, Anglo-American philosophy entered upon a period of self-imposed isolation from developments that might have pointed a way beyond some of its more intractable dilemmas.

Now at last there are signs – for instance, in the recent work of Michael Dummett – that analytic philosophers are coming to question this received view, especially as concerns the orthodox charge (handed down by Frege and Ryle) that Husserl's project was really just a form of naive psychologism decked out in spurious 'transcendental' colours.[104] Also there is a growing recognition that philosophy of science in the broadly analytic mode may yet have something of importance to learn from the critical-rationalist approach of thinkers such as Bachelard and his student Georges Canguilhem.[105] Perhaps the next stage in this process will be a reading of Derrida's early texts that acknowledges both their analytic rigor and their bearing on issues, such as those discussed here, that have been either ignored or inadequately treated by philosophers in the mainstream Anglo-American line of descent.

Notes

1 Frege (1972); Ryle (1971a, 1971b, 1971c). See also Follesdal (1994) and Tito (1990).
2 See especially Husserl (1970a, 1970b, 1973a, 1973b, 1975).
3 Husserl (1969).
4 Dummett (1993).
5 Quine (1961).
6 See Quine (1971a, 1971b).
7 See, for instance, the various discussions in Strawson (1967).
8 See Marcus (1971), p. 49.
9 Ibid., p. 51.
10 Ibid., p. 47.
11 Ibid., p. 48.
12 See, for instance, – from a range of philosophic viewpoints – Mulhall (1990), Okrent (1988), Rorty (1991a, 1991b) and Rouse (1987). For more extensive discussion, see Norris (1997a, 1997b, 1997c).

13 Quine (1961), p. 41.
14 Ibid., p. 42.
15 Ibid.
16 Ibid., p. 37.
17 Ibid., p. 43.
18 Ibid., p. 44.
19 Ibid., p. 45.
20 See especially the essays collected in Rorty (1991a).
21 Rorty (1986). For a critical discussion of Rorty's arguments in this context, see Child (1994).
22 Davidson (1980).
23 See, for instance, Churchland (1979, 1980).
24 Davidson (1984a).
25 Rorty (1986), p. 353.
26 Ibid., p. 341.
27 Ibid., pp. 347–8.
28 Ibid., p. 353.
29 See Quine (1961); also Quine (1960, 1969).
30 Davidson (1984b), p. 198.
31 See entries under note 12, above; also Malpas (1992).
32 See notes 12 and 31, above; Dreyfus (1991).
33 Evnine (1991), p. 175; cited by Child (1994), p. 121.
34 See especially Papineau (1987, 1993).
35 Davidson (1984b).
36 Rorty (1986).
37 See note 12 above; also Rorty (1980).
38 Child (1994), p. 132.
39 Davidson (1980), p. 241.
40 Quine (1961), p. 46.
41 See especially Bachelard (1933, 1937, 1938, 1949, 1951, 1953, 1969, 1984). See also Gaukroger (1986), Gutting (1987), Lecourt (1975) and Tiles (1984).
42 Quine (1961), p. 46.
43 Kuhn (1970).
44 Quine (1970).
45 For a range of views on this topic, see Beltrametti and van Fraassen (1981), Garden (1989), Gibbins (1987), Haack (1974), Mittelstaedt (1994) and Putnam (1979).
46 See, for instance, Fine (1986), Jammer (1974), Krips (1987), Maudlin (1993) and Redhead (1987).
47 See Rorty (1991a).
48 See Schrödinger (1967); also Gribbin (1984).
49 See notes 46 and 48 above; also Audi (1973), Bell (1987), Forrest (1988), Rae (1986), Reichenbach (1944) and Squires (1994).
50 See especially Bohr (1934, 1958); also Folse (1985), Honner (1987) and Murdoch (1987).
51 See Bohr (1949); also Agazzi, (1997), Albert, (1994), Bohm and Hiley (1993), Cushing (1994), Cushing and McMullin, (1989) and Popper (1982).
52 See, for instance, Luntley (1988), Wright (1987) and Tennant (1987).
53 See especially Dummett (1978a).

54 For further discussion from a range of philosophical standpoints, see Devitt (1986), Farrell (1996), Kirkham (1989), Norris (1997a), Platts (1980), Vision (1988), Williamson (1988), Weir (1986) and Winkler (1985).
55 Dummett (1977, 1978a).
56 See Dummett (1978b).
57 See note 41, above.
58 See especially Bachelard (1938).
59 See especially Tiles (1984).
60 Lakatos (1978), p. 102.
61 See Coffa (1991).
62 Derrida (1978a, 1978b). See also Derrida (1973).
63 Derrida (1982), esp. pp. 258–66.
64 Derrida (1978a), pp. 158–9.
65 Derrida (1978b), p. 152.
66 See, for instance, Feyerabend (1975, 1978).
67 See entries under notes 5, 20, 21, 24, 29 and 43 above; also Putnam (1987, 1988, 1990).
68 Davidson (1984b), p. 198.
69 See note 67, above; also various contributions to Clark and Hale (1993).
70 For further discussion see Norris (1997a, 1997c).
71 See note 67, above; also Putnam (1995).
72 Kuhn (1970), p. 126.
73 See, for instance, Carnap (1956, 1967, 1974).
74 See especially Husserl (1970a).
75 Derrida (1978a), p. 158.
76 Ibid., p. 162.
77 See, for instance, the essays collected in Ayer (1959); also Hanfling (1981) and Rescher (1985).
78 See Reichenbach (1938).
79 For examples of the 'strong-sociological' approach, see Barnes (1985), Bloor (1976), Brannigan (1981), Collins (1985), Collins and Pinch (1993), Fuller*s* (1989), Latour and Woolgar (1979), Pickering (1992), Shapin and Schaffer (1985) and Woolgar (1988).
80 Derrida (1978a), p. 158.
81 Derrida (1978c).
82 Derrida (1978a), p. 160.
83 Ibid., p. 162.
84 Penrose (1995), p. 65.
85 Ibid., p. 111.
86 Ibid., p. 72.
87 Ibid., p. 64.
88 See Wittgenstein (1967, 1969, 1976); also Kripke (1982) and – for a sharply opposed interpretation – Wright (1980).
89 See, for instance, Bloor (1983) and Phillips (1977).
90 Penrose (1995), p. 65.
91 See Lyotard (1984).
92 See notes 62 and 63, above; also Derrida (1976, 1981). Some of the best commentary on these and related aspects of his work may be found in Gasché (1986).
93 Derrida (1978a), p. 159.

94 De Saussure (1983). For further texts and commentary, see Hawkes (1977), Lane (1970), Macksey and Donato (1972), Pavel (1990) and Sturrock (1979).
95 See especially Ricoeur (1974) and Merleau-Ponty (1968).
96 Derrida (1978a), p. 167.
97 Foucault (1970).
98 See, for instance, Foucault (1973).
99 Norris (1997d).
100 See entries under note 79 above.
101 Penrose (1995), p. 65.
102 Derrida (1978a), p. 166; passages cited from Husserl (1973a), pp. 139 and 156.
103 See especially Carnap (1959).
104 See Dummett (1993).
105 See, for instance, Canguilhem (1968, 1969, 1988).

References

Agazzi, Evadro (ed.) (1997), *Realism and Quantum Physics*, Amsterdam and Atlanta: Rodopi.

Albert, David Z. (994), 'Bohm's Alternative to Quantum Mechanics', *Scientific American*, No. 270 (1 May), pp. 50–63.

Audi, Michael (1973), *The Interpretation of Quantum Mechanics*, Chicago: University of Chicago Press.

Ayer, A.J. (ed.) (1959), *Logical Positivism*, New York: Free Press.

Bachelard, Gaston (1933), *Les intuitions atomistiques*, Paris: Boivin.

Bachelard, Gaston (1937), *L'expérience de l'espace dans la physique contemporaine*, Paris: Presses Universitaires de France.

Bachelard, Gaston (1938), *La formation de l'esprit scientifique*, Paris: Vrin.

Bachelard, Gaston (1949), *Le rationalisme appliqué*, Paris: PUF.

Bachelard, Gaston (1951), *L'activité rationaliste de la physique contemporaine*. Paris: PUF.

Bachelard, Gaston (1953), *Le matérialisme rationnel*, Paris: PUF.

Bachelard, Gaston (1969), *The Philosophy of No*, trans. G.C. Waterston, New York: Orion Press.

Bachelard, Gaston (1984), *The New Scientific Spirit*, trans. Arthur Goldhammer, Boston: Beacon Press.

Barnes, Barry (1985), *About Science*, Oxford: Basil Blackwell.

Beltrametti, Enrico G. and van Fraassen, Bas C. (1981), *Current Issues in Quantum Logic*, New York: Plenum.

Bell, J.S. (1987), *Speakable and Unspeakable in Quantum Mechanics: Collected Papers on Quantum Philosophy*, Cambridge: Cambridge University Press.

Bloor, David (1976), *Knowledge and Social Imagery*, London: Routledge and Kegan Paul.

Bloor, David (1983), *Wittgenstein: A Social Theory of Knowledge*, New York: Columbia University Press.

Bohm, David and Hiley, Basil J. (1993), *The Undivided Universe: An Ontological Interpretation of Quantum Theory*, London: Routledge.

Bohr, Niels (1934), *Atomic Theory and the Description of Nature*, Cambridge: Cambridge University Press.

Bohr, Niels (1949), 'Discussion with Einstein on Epistemological Problems in Atomic Physics', in Paul A. Schilpp (ed.), *Albert Einstein: Philosopher-scientist*, La Salle, IL: Open Court, pp. 201–41.

Bohr, Niels (1958), *Atomic Physics and Human Knowledge*, New York: Wiley.

Brannigan, Augustine (1981), *The Social Basis of Scientific Discoveries*, Cambridge: Cambridge University Press.

Canguilhem, Georges (1968), *Etudes d'histoire et de la philosophie des sciences*, Paris: Vrin.

Canguilhem, Georges (1969), *La connaissance de la vie*, 2nd edn, Paris: Vrin.

Canguilhem, Georges (1988), *Ideology and Rationality in the History of the Life Sciences*, trans. Arthur Goldhammer, Cambridge, MA: MIT Press.

Carnap, Rudolf (1956), *Meaning and Necessity*, Chicago: University of Chicago Press.

Carnap, Rudolf (1959), 'The Elimination of Metaphysics through Logical Analysis of Language', in A.J. Ayer (ed.), *Logical Positivism*, New York: Free Press, pp. 60–81.

Carnap, Rudolf (1967), *The Logical Structure of the World and Pseudoproblems in Philosophy*, trans. R. George, Berkeley and Los Angeles: University of California Press.

Carnap, Rudolf (1974), *An Introduction to the Philosophy of Science*, New York: Basic Books.

Child, William (1994), *Causality, Interpretation and the Mind*, Oxford: Clarendon Press.

Churchland, Paul (1979), *Scientific Realism and the Plasticity of Mind*, Cambridge: Cambridge University Press.

Churchland, Paul (1980), *Matter and Consciousness*, Cambridge, MA: MIT Press.

Clark, Peter and Hale, Bob (eds) (1993), *Reading Putnam*, Oxford: Basil Blackwell.

Coffa, J. Alberto (1991), *The Semantic Tradition from Kant to Carnap: To the Finland Station*, Cambridge: Cambridge University Press.

Collins, Harry (1985), *Changing Order: Replication and Induction in Scientific Practice*, Chicago: University of Chicago Press.

Collins, Harry and Pinch, Trevor (1993), *The Golem: What Everyone Should Know about Science*, Cambridge University Press.

Cushing, James T. (1994), *Quantum Mechanics: Historical Contingency and the Copenhagen Hegemony*, Chicago: University of Chicago Press.

Cushing, James T. and McMullin, Ernan (eds) (1989), *Philosophical Consequences of Quantum Theory: Reflections on Bell's Theorem*, Notre Dame, IN: University of Notre Dame Press.

Davidson, Donald (1980), *Essays on Actions and Events*, London: Oxford University Press.

Davidson, Donald (1984a), *Inquiries Into Truth and Interpretation*, London: Oxford University Press.

Davidson, Donald (1984b), 'On the Very Idea of a Conceptual Scheme', in *Inquiries into Truth and Interpretation*, London: Oxford University Press, pp. 183–98.

Derrida, Jacques (1973), *'Speech and Phenomena' and Other Essays on Husserl's Theory of Signs*, trans. David B. Allison, Evanston, IL: Northwestern University Press.

Derrida, Jacques (1976), *Of Grammatology*, trans. G.C. Spivak, Baltimore, MD: Johns Hopkins University Press.

Derrida, Jacques (1978a), '"Genesis and Structure" and Phenomenology', in *Writing and Difference*, trans. Alan Bass, London: Routledge and Kegan Paul, pp. 154–68.

Derrida, Jacques (1978b), 'Violence and Metaphysics: An Essay on the Thought of Emmanuel Levinas', in *Writing and Difference*, trans. Alan Bass, London: Routledge and Kegan Paul, pp. 79–153.

Derrida, Jacques (1978c), *Edmund Husserl's 'Origin of Geometry': An Introduction*, trans. John P. Leavey, Pittsburgh, PA: Duquesne University Press.

Derrida, Jacques (1981), *Dissemination*, trans. Barbara Johnson, London: Athlone Press.

Derrida, Jacques (1982), 'White Mythology: Metaphor in the Text of Philosophy', in *Margins of Philosophy*, trans. Alan Bass, Chicago: University of Chicago Press pp. 207–71.

De Saussure, Ferdinand (1983), *Course in General Linguistics*, trans. Roy Harris, London: Duckworth.

Dreyfus, Hubert L. (1991), *Being-in-the-World: A Commentary on Heidegger's Being and Time, Division One*, Cambridge, MA: MIT Press.

Devitt, Michael (1986), *Realism and Truth*, 2nd edn, Oxford: Basil Blackwell.

Dummett, Michael (1977), *Elements of Intuitionism*, London: Oxford University Press.

Dummett, Michael (1978a), *Truth and Other Enigmas*, London: Duckworth.

Dummett, Michael (1978b), 'The Significance of Quine's Indeterminacy Thesis', in *Truth and Other Enigmas*, London: Duckworth.

Dummett, Michael (1993), *The Origins of Analytic Philosophy*, London: Duckworth.

Evnine, Simon (1991), *Donald Davidson*, Oxford: Polity Press.

Farrell, Frank B. (1996), *Subjectivity, Realism and Postmodernism: The Recovery of the World in Recent Philosophy*, Cambridge: Cambridge University Press.

Feyerabend, Paul K. (1975), *Against Method*, London: New Left Books.

Feyerabend, Paul K. (1978), *Science in a Free Society*, London: New Left Books.

Fine, Arthur (1986), *The Shaky Game: Einstein, Realism, and Quantum Theory*, Chicago: University of Chicago Press.

Follesdal, Dagfinn (1994), 'Husserl and Frege: A Contribution to Elucidating the Origins of Phenomenological Philosophy', in Leila Haaparanta (ed.), *Mind, Meaning and Mathematics: Essays on the Philosophical Views of Husserl and Frege*, Dordrecht and Boston: Kluwer, pp. 3–47.

Folse, Henry J. (1985), *The Philosophy of Niels Bohr: The Framework of Complementarity*, Amsterdam: North-Holland.

Forrest, Peter (1988), *Quantum Metaphysics*, Oxford: Basil Blackwell.

Foucault, Michel (1970), *The Order of Things: An Archaeology of the Human Sciences*, trans. Alan Sheridan, London: Tavistock.

Foucault, Michel (1973), *The Archaeology of Knowledge*, trans Alan Sheridan, London: Tavistock.

Frege, Gottlob (1972), 'Review of Edmund Husserl's *Philosophie der Arithmetik*', trans. E.-H.W. Kluge, *Mind*, Vol. LXXXI, pp. 321–37.

Fuller, Steve (1989), *Philosophy of Science and its Discontents*, Boulder, CO: Westview Press.

Garden, Rachel Wallace (1989), *Modern Logic and Quantum Mechanics*, Bristol: A. Hilger.

Gasché, Rodolphe (1986), *The Tain of the Mirror: Derrida and the Philosophy of Reflection*, Cambridge, MA: Harvard University Press.

Gaukroger, Stephen W. (1986), 'Bachelard and the Problem of Epistemological Analysis', *Studies in History and Philosophy of Science*, Vol. XVII, pp. 189–244.

Gibbins, Peter (1987), *Particles and Paradoxes: The Limits of Quantum Logic*, Cambridge: Cambridge University Press.

Gribbin, John (1984), *In Search of Schrödinger's Cat: Quantum Physics and Reality*, New York: Bantam Books.

Gutting, Gary (1987), 'Gaston Bachelard's Philosophy of Science', *International Studies in the Philosophy of Science*, Vol. II, pp. 55–71.

Haack, Susan (1974), *Deviant Logic: Some Philosophical Issues*, Cambridge University Press.

Hanfling, Oswald (ed.) (1981), *Essential Readings in Logical Positivism*, Oxford: Basil Blackwell.

Hawkes, Terence (1977), *Structuralism and Semiotics*, London: Methuen.

Honner, John (1987), *The Description of Nature: Niels Bohr and the Philosophy of Quantum Physics*, Cambridge: Cambridge University Press.

Husserl, Edmund (1969), *Formal and Transcendental Logic*, trans. Dorion Cairns, The Hague: Martinus Nijhoff.

Husserl, Edmund (1970a), *The Crisis of European Sciences and Transcendental Phenomenology: An Introduction to Phenomenological Philosophy*, trans. David Carr, Evanston, IL: Northwestern University Press.

Husserl, Edmund (1970b), *Logical Investigations*, 2 vols, trans. J.N. Findlay, New York: Humanities Press.

Husserl, Edmund (1973a), *Cartesian Meditations: An Introduction to Phenomenology*, trans. Dorion Cairns, The Hague: Martinus Nijhoff.

Husserl, Edmund (1973b), *Experience and Judgement: Investigations in a Genealogy of Logic*, trans. James S. Churchill and Karl Ameriks, Evanston, IL: Northwestern University Press.

Husserl, Edmund (1975), *Ideas*: *General Introduction to Pure Phenomenology*, trans. W.R. Boyce Gibson, London: Collier Macmillan.

Jammer, Max (1974), *Philosophy of Quantum Mechanics: The Interpretations of Quantum Mechanics in Historical Perspective*, New York: Wiley.

Kirkham, Richard L. (1989), 'What Dummett Says About Truth and Linguistic Competence', *Mind*, Vol. XCVIII, pp. 207–24.

Kripke, Saul (1982), *Wittgenstein on Rules and Private Language*, Oxford: Basil Blackwell.

Krips, Henry (1987), *The Metaphysics of Quantum Theory*, Oxford: Clarendon Press.

Kuhn, Thomas S. (1970), *The Structure of Scientific Revolutions*, 2nd edn, Chicago: University of Chicago Press.

Lakatos, Imré (1978), *Philosophical Papers*, Vol. 1: *The Methodology of Scientific Research Programmes*, Cambridge: Cambridge University Press.

Lane, Michael (ed.) (1970), *Structuralism:: A Reader*, London: Allen Lane.

Latour, Bruno and Woolgar, Steve (1979), *Laboratory Life: The Social Construction of Scientific Facts*, London: Sage.

Lecourt, Dominique (1975), *Marxism and Epistemology: Bachelard, Canguilhem, Foucault*, trans. Ben Brewster, London: New Left Books.

Luntley, Michael (1988), *Language, Logic and Experience*: *The Case for Anti-realism*, London: Duckworth.

Lyotard, Jean-François (1984), *The Postmodern Condition*, trans. Geoff Bennington and Brian Massumi, Manchester: Manchester University Press.

Macksey, Richard and Donato, Eugenio (eds) (1972), *The Structuralist Controversy: The Languages of Criticism and the Sciences of Man*, Baltimore, MD: Johns Hopkins University Press.

Malpas, J.E. (1992), *Donald Davidson and the Mirror of Meaning*, Cambridge: Cambridge University Press.

Marcus, Ruth B. (1971), 'Extensionality', in Leonard Linsky (ed.), *Reference and Modality*, London: Oxford University Press, pp. 44–51.

Maudlin, Tim (1993), *Quantum Nonlocality and Relativity: Metaphysical Implications of Modern Science*, Oxford: Basil Blackwell.

Merleau-Ponty, Maurice (1968), *The Visible and the Invisible*, ed. Claude Lefort, trans. Alphonso Lingis, Evanston, IL: Northwestern University Press.
Mittelstaedt, Peter (1994), *Quantum Logic*, Princeton, NJ: Princeton University Press.
Mulhall, Stephen (1990), *On Being in the World: Wittgenstein and Heidegger on Seeing Aspects*, London: Routledge.
Murdoch, Dugald (1987), *Niels Bohr's Philosophy of Physics*, Cambridge: Cambridge University Press.
Norris, Christopher (1997a), *Resources of Realism: Prospects for 'Post-analytic' Philosophy*, London: Macmillan.
Norris, Christopher (1997b), *New Idols of the Cave: On the Limits of Anti-realism*, Manchester: Manchester University Press.
Norris, Christopher (1997c), *Against Relativism: Philosophy of Science, Deconstruction and Critical Theory*, Oxford: Basil Blackwell.
Norris, Christopher (1997d), 'Ontological Relativity and Meaning-variance: A Critical-constructive Review', in *Against Relativism: Philosophy of Science, Deconstruction and Critical Theory*, Oxford: Basil Blackwell, pp. 66–100.
Okrent, Mark (1988), *Heidegger's Pragmatism: Understanding, Being, and the Critique of Metaphysics*, Ithaca, NY: Cornell University Press.
Papineau, David (1987), *Reality and Representation*, Oxford: Basil Blackwell.
Papineau, David (1993), *Philosophical Naturalism*, Oxford: Basil Blackwell.
Pavel, Thomas (1990), *The Feud of Language: A History of Structuralist Thought*, Oxford: Basil Blackwell.
Penrose, Roger (1995), *Shadows of the Mind: A Search for the Missing Science of Consciousness*, London: Vintage.
Phillips, Derek L. (1977), *Wittgenstein and Scientific Knowledge: A Sociological Perspective*, London: Macmillan.
Pickering, Andrew (1992), *Science as Practice and Culture*, Chicago: University of Chicago Press.
Platts, Mark (ed.) (1980), *Reference, Truth and Reality: Essays on the Philosophy of Language*, London: Routledge and Kegan Paul.
Popper, Karl (1982), *Quantum Theory and the Schism in Physics*, London: Hutchinson.
Putnam, Hilary (1979), *Mathematics, Matter and Method*, Cambridge: Cambridge University Press.
Putnam, Hilary (1987), *The Many Faces of Realism*, La Salle, IL: Open Court.
Putnam, Hilary (1988), *Representation and Reality*, Cambridge: Cambridge University Press.
Putnam, Hilary (1990), *Realism With a Human Face*, Cambridge, MA: Harvard University Press.
Putnam, Hilary (1995), *Pragmatism: An Open Question*, Oxford: Basil Blackwell.
Quine, W.V. (1960), *Word and Object*, Cambridge, MA: MIT Press.
Quine, W.V. (1961), 'Two Dogmas of Empiricism', in *From a Logical Point of View*, 2nd edn, Cambridge, MA: Harvard University Press, pp. 20–46.
Quine, W.V. (1969), *Ontological Relativity and Other Essays*, New York: Columbia University Press.
Quine, W.V. (1970), *Philosophy of Logic*, Englewood Cliffs, NJ: Prentice-Hall.
Quine, W.V. (1971a), 'Reference and Modality', in Leonard Linsky (ed.), *Reference and Modality*, London: Oxford University Press, pp. 17–34.
Quine, W.V. (1971b), 'Quantifiers and Propositional Attitudes', in Leonard Linsky (ed.), *Reference and Modality*, London: Oxford University Press, pp. 101–11.

Rae, Alastair I.M (1986), *Quantum Physics: Illusion or Reality?*, Cambridge: Cambridge University Press.
Redhead, Michael (1987), *Incompleteness, Nonlocality and Realism: A Prolegomenon to the Philosophy of Quantum Mechanics*, Oxford: Clarendon Press.
Reichenbach, Hans (1938), *Experience and Prediction*, Chicago: University of Chicago Press.
Reichenbach, Hans (1944), *Philosophic Foundations of Quantum Mechanics*, Berkeley and Los Angeles: University of California Press.
Rescher, Nicholas (ed.) (1985), *The Heritage of Logical Positivism*, Lanham, MD: University Press of America.
Ricoeur, Paul (1974), *The Conflict of Interpretations: Essays in Hermeneutics*, ed. Don Ihde, Evanston, IL: Northwestern University Press.
Rorty, Richard (1980), *Philosophy and the Mirror of Nature*, Oxford: Basil Blackwell.
Rorty, Richard (1986), 'Pragmatism, Davidson, and Truth', in Ernest LePore (ed.), *Truth and Interpretation: Perspectives on the Work of Donald Davidson*, Oxford: Basil Blackwell, pp. 333–55.
Rorty, Richard (1991a), *Objectivity, Relativism, and Truth*, Cambridge: Cambridge University Press.
Rorty, Richard (1991b), *Essays on Heidegger and Others*, Cambridge: Cambridge University Press.
Rouse, Joseph (1987), *Knowledge and Power: Toward a Political Philosophy of Science*, Ithaca, NY: Cornell University Press.
Ryle, Gilbert (1971a), 'Phenomenology', in *Collected Papers*, Vol. 1, London: Hutchinson, pp. 167–78.
Ryle, Gilbert (1971b), 'Review of Martin Farber, *The Foundations of Phenomenology*', in *Collected Papers*, Vol. 1, London: Hutchinson, pp. 215–24.
Ryle, Gilbert (1971c), 'Phenomenology versus *The Concept of Mind*', in *Collected Papers*, Vol. 1, London: Hutchinson, pp. 179–96.
Schrödinger, Erwin (1967), *Letters on Wave Mechanics*, New York: Philosophical Library.
Shapin, Steven and Schaffer, Simon (1985), *Leviathan and the Air-pump: Hobbes, Boyle, and the Experimental Life*, Princeton, NJ: Princeton University Press.
Squires, Euan (1994), *The Mystery of the Quantum World*, 2nd edn, Bristol: Institute of Physics Publishing.
Strawson, P.F. (ed.) (1967), *Philosophical Logic*, London: Oxford University Press.
Sturrock, John (1979), *Structuralism and Since*, London: Oxford University Press.
Tennant, N. (1987), *Anti-realism and Logic*, Oxford: Clarendon Press.
Tiles, Mary (1984), *Bachelard: Science and Objectivity*, Cambridge: Cambridge University Press.
Tito, Johanna Maria (1990), *Logic in the Husserlian Context*, Evanston, IL: Northwestern University Press.
Vision, Gerald (1988), *Modern Anti-realism and Manufactured Truth*, London: Routledge.
Weir, Alan (1986), 'Dummett on Meaning and Classical Logic', *Mind*, Vol. XCV, pp. 465–77.
Williamson, Timothy (1988), 'Knowability and Constructivism: The Logic of Anti-realism', *Philosophical Quarterly*, Vol. XLVIII , pp. 422–32.
Winkler, Kenneth P. (1985), 'Skepticism and Anti-realism', *Mind*, Vol. XCIV, pp. 36–52.
Wittgenstein, Ludwig (1967), *Philosophical Investigations*, trans. G.E.M. Anscombe, Oxford: Basil Blackwell.

Wittgenstein, Ludwig (1969), *On Certainty*, trans. D. Paul and G.E.M. Anscombe, Oxford: Basil Blackwell.
Wittgenstein, Ludwig (1976), *Wittgenstein's Lectures on the Foundations of Mathematics*, ed. Cora Diamond, Chicago: University of Chicago Press.
Woolgar, Steve (1988), *Science: The Very Idea*, London: Tavistock.
Wright, Crispin (1980), *Wittgenstein on the Foundations of Mathematics*, Cambridge, MA: Harvard University Press.
Wright, Crispin (1987), *Realism, Meaning and Truth*, Oxford: Basil Blackwell.

Chapter 3

Science, History and Philosophy in Kant and Hegel

Angelica Nuzzo

Reason, and more precisely human reason, is the power that generates science and in doing philosophy reflects upon its own activity and accomplishments. This general claim may be taken as framing both Kant's and Hegel's consideration of what science is and ought to be. In their account, *Wissenschaft* is identified, first and foremost, through its rational source, since the source determines both science's methodology, its problems and procedures, and the extent and legitimacy of its practices. Under this premise, the question of what counts as *scientific* experience and of how scientific *experience* is produced out of that source is the problem that both Kant and Hegel immediately link to the determination of the nature and structures of rationality. While for Kant reason is still conceived as a subjective 'faculty' or 'capacity' (*Vermögen*) whose correct and justified exercise or 'use' (*Gebrauch*) is responsible for the institution of scientific cognition, Hegel identifies reason with the actual realization of rational activity: rationality is not just a form of cognition (theoretical or practical), but an immanent and objective dimension of reality itself; reason is not a set of pre-conditions to knowledge, but the dimension of reality in its actual processuality. In this sense, for Hegel, reason is indeed *ab-soluta* from all dependency on a finite human subject. Reason becomes the dimension of the 'absolute method' that constitutes *Wissenschaft* as such. Despite this fundamental difference, however, both for Kant and for Hegel the philosophical account of the structures of rationality must take on the form of a complete and concluded system of fundamental forms and functions that, as categories of the understanding in transcendental logic and as pure forms of thinking in speculative logic, normatively permeate all successful endeavor in doing science.

The fundamental question that I want to address is then the following: How can such systematic, logic-oriented programs as Kant's and Hegel's ever allow for an intervention of the condition of history in science that were more than extrinsic and ultimately irrelevant? With the discussion of this issue in Kant's and Hegel's philosophy, I want to show the way in which a systematic theory of rationality that finds its ground in logic can indeed make possible an articulated view of historicity as the intrinsic dimension of reason itself – of its activity and objective reality at the same time. In this perspective, which eventually corresponds to Hegel's dialectical position, the relation between history and science becomes the constitutive and necessary relation that is in charge of the process that institutes scientific rationality itself, and thereby even grounds the possibility of doing 'history of science.'

In the following considerations, the crucial issue concerning the ways in which history and historicity are differently implied in Kant's and Hegel's conceptions of science and scientific knowledge is raised in connection with their views of reason and rationality. The issue of the scientific character proper to the exact sciences is not always explicitly distinguished from the issue of the scientific character claimed for philosophy itself. The central problem regards the idea of 'history' that results, respectively, from Kant's and Hegel's notions of science and philosophy as products of reason's own activity. I analyze this general issue with regard to the following question: *at what stage* in the constitution of science and scientific rationality which is attempted, respectively, by Kant's critical philosophy and Hegel's dialectic, does the condition of 'history' intervene, and how does this condition affect the constitution of rationality and rational theories (scientific as well as philosophical)?

Put in another way, the question that presently interests me can be seen as an instance of the problem raised by the apparently impossible relation between the invariance or eternity of logical truth and the constitutive mutability of time and our experience of that truth.[1] In the Introductions with which Hegel used to open his *Lectures on the History of Philosophy* held every year between 1819 and 1830, he never ceased to point out the importance of this paradox for the very possibility of conceiving the historicity proper to reason.[2] While that paradox or contradiction still produces an unresolved tension in Kant's critical philosophy and, in particular, in the relation between the results of the *Critique of Pure Reason* and the later *Critique of Judgment*, for Hegel that contradiction becomes the generative source of the idea of history in so far as speculative philosophy takes into account the specific history or historicity inherent to scientific thinking. For Hegel, the idea of history plays itself out precisely in the intersection between logical invariance and structural processuality. At this juncture, Hegel constitutes the idea of the dialectic as 'method.'

Accordingly, the main issue of this essay can be expressed as follows: how do the two extremes of truth and time, reason and history meet in Kant's and Hegel's view science? The thesis that allows both Kant and Hegel to insert – albeit in different ways and with a different consciousness – the condition of history at the very heart of scientific rationality regards not directly the idea of truth, but rather, respectively, the possibility of *error* in our quest for truth (Kant's unavoidable dialectical illusion) and in the constitution of the complete system of all truth (Hegel's 'absolute knowing' at the end of the phenomenological process through which consciousness reaches truth by experiencing the complete set of all possible errors). The possibility for the achieved truth to be 'open' to further completion and modification, and hence its genuine 'historicity,' ultimately depends upon truth's relation to error.

1 In this essay I analyze, first, the general structure and the different implications of the claim that both thinking and the reality which is object and product of our thinking are inherently historical. Such a claim, in a more precise formulation, can be seen as the final accomplishment of Hegel's dialectical theory of rationality. The first task is to see how Kant's program of a transcendental philosophy or the crucial critical question regarding the possibility of science and scientific experience relates to that claim. Kant's position is compared with and contrasted to the specific form of historicity that Hegel attributes to reason as *Geist*. I will limit my considerations to a discussion of the relation between Hegel's notion of 'objective thinking' as the key to the development of the logical process, and the idea of *Geist* presented in the conclusion of the *Phenomenology of Spirit*. Thereby I want to underscore the import of the phenomenological component of Hegel's dialectic for the thesis of an inherent historicity of science.

2 The second part of the essay is dedicated to the discussion of the idea of 'history of pure reason' that emerges at the very end of the Transcendental Doctrine of Method of Kant's *Critique of Pure Reason*. At stake herein is the issue of the relation between the historicity inherent to philosophy and that proper to the exact sciences, mathematics in particular. While both philosophical and mathematical cognition are defined as 'rational cognition' (*Vernunfterkenntis*), the way in which reason is at work and is subject to error in the two disciplines is radically different.

1 History, Historicity and the Problem of Science

In discussing the thesis of the intrinsic historicity of scientific (and philosophical) thinking, we have to keep in mind that the German terminology (like the Latin one) accurately distinguishes between *Historie* (*historia rerum gestarum*) and *Geschichte* (*res gestae*). In addition, it is relevant to note that the term *Geschichtlichkeit* which is commonly translated as 'historicity,' is relatively recent in the German vocabulary as a technical philosophical term. In Kant's writings, the former distinction is not yet thematized, and the word *Geschichtlichkeit* never appears. Hegel dedicates a complex reflection to the need for differentiating, conceptually as well as terminologically, the diverse meanings of the notion of history. The well-known distinction between 'original,' 'reflected' and 'philosophical' history opens Hegel's courses on the philosophy of history.[3] Moreover, even within the notion of philosophical history further distinctions are in order. In his Nürnberg lectures held between 1808 and 1816, and published posthumously under the title *Philosophische Propedeutik*, for example, Hegel discusses the different figures of *begriffene Geschichte* (which is already mentioned at the end of the *Phenomenology of Spirit*). Thereby he distinguishes *philosophische Geschichte*, *philosophische Ansicht der Geschichte* and *historische Geschichte*.[4] Hegel is probably the first to use the term *Geschichtlichkeit* which occurs in two passages both to be found in the *Lectures on the History of Philosophy*.[5] Through Hegel, the term passes on to the *Hegel-Schule* (in particular to Henrich Heine and Karl Rosenkranz), and then, through Rudolph Haym, reaches Dilthey and York, who eventually provide for it the constellation of technical meanings that from now on will characterize the use of the term *Geschichtlichkeit* in important developments of German philosophy.

While for Dilthey and York the term 'historicity' has the specific function of designating the modality proper to human existence, for Hegel it simply refers to the commonsense idea according to which that of which historicity is predicated *has a history*, *is in history*, and displays 'historical' existence. Obviously, all these translations of the term 'historicity' do not explain the term itself. They only contribute to present the constellation of problems covered by it. Historicity is predicated, according to Hegel, on thinking, reason and truth; it never designates an existential modality restricted to the human being or alluding to its finitude. In addition, historicity is by no means a notion which depends upon a primary role played by time.

The relevant question, given these terminological specifications, is now the following: What is the 'historicity' proper to science or constitutive of scientific thinking and of the practices that institute the reality of science?

The gesture of going back to the context that hosts the 'historical' origin of the term 'historicity' in Hegel's philosophy may help to shed some light on this issue.

The question of science's historicity must be articulated in two distinct claims.[6] At stake is the 'genetic' definition of the idea of science, not the conditions for doing a 'history of science.' For, both Kant and Hegel are not primarily concerned with science as an historically given set of practices and theories. To the extent that science is, for them, the foremost activity of rationality itself, that is, that in which rationality manifests its power and the full range of its capabilities, science poses to philosophy the problem of the justification of its procedures. Thus, it is at this level that the question of the historicity of science must be addressed:

1. First, I want to set Kant's and Hegel's philosophical theories in relation with the general claim that *thinking* is constitutively and inherently historical. With regard to this thesis, the question must be addressed of *where* must the historicity of thinking be located, or *what* makes the historicity of thinking: is it its inner logical structure, its use and exercise through different practices, its external appearance in time, the progressive transformation of its problems and tasks?
2. The second general claim is that the *reality* to which scientific cognition is oriented and which science, together with all other human activities, contributes to form, is itself inherently historical. In this case as well, reality's 'historicity' must be specified with regard to the ontological layers that history may contribute to shape, that is, to the distinction between contingent events and necessary structures, appearances and substance, reality as a given and reality as a construction. The question of *what is historical in reality and what is real in history* deeply concerns, albeit in different ways and with different consequences, both Kant and Hegel.

In the present connection, however, my aim is not to provide a discussion of each of these claims taken separately. What interests me is rather the relation between them and the point at which they may intersect. For this point would disclose a significant meaning for the notion of 'historicity.' Such a move may be seen as a different approach to the issue of the general conflict between Kant's 'dualism' and Hegel's 'monism' that, alternatively, either separates or unifies thinking and reality, logical invariance and historical mutability. The question that I want to pose will contribute to present this general issue in a more complex and articulated way.

1.1 Kant's Critical Question and the Possibility of Physics

In order to correctly appreciate whether and how the epistemological theory of the *Critique of Pure Reason* hinges upon the assumption of the validity of Newtonian physics,[7] that is, of an historically specific stage of the development of science, we need to clarify what is the *critical* and *transcendental* question that Kant asks in his 1781/87 work. Kant's *critical* problem concerns the *possibility of physics* and the *possibility of (scientific) experience*; it addresses the formal conditions of the possibility of experience as the order and connection of appearances that for us constitute 'nature *in general*.' Accordingly, the *Critique of Pure Reason* is an inquiry into the structures of experience proper to physics *in general* and its *universal* laws. To this extent, the critical and transcendental question of the first *Critique* differs radically both from the *metaphysical* problem that Kant raises in the *Metaphysical Foundations of Natural Science*, where he investigates the *particular* empirical laws of physics, and from the new *transcendental* question addressed by the *Critique of Judgment* with regard to the myriad of particular empirical laws of natural sciences such as biology, anatomy, geology, and in general all those sciences that make use of taxonomy.[8]

Since mathematical and physical sciences *do exist as a matter of fact*, and produce indeed objective cognition of the physical world, the critical task is to find the ground for the objective validity of science. This ground lies in the human cognitive faculty and its functions. Thus, to provide the theoretical justification of physics means to examine the ways in which – and the principles according to which – human reason investigates nature. Thereby, the critical question of Kant's philosophy translates the *historical* 'fact' that natural sciences exist in certain theories and laws, into the *transcendental* conditions for the institution of that historical fact in its most general form through human reason. The historical reality of science is produced by the transcendental structures of human reason. Under this assumption, the issue of the 'objective reality' of scientific judgments is translated into the problem of their 'objective validity.' Hence, with Kant's critical question, the issue of science's historicity shifts from 'reality' to 'validity,' namely to the level of the transcendental conditions of human rationality. What role does history play at this level?

Kant's critical investigation hinges upon the crucial notion of 'experience,' which he formulates according to the terms of his famous 'Copernican revolution.' With his notion of 'experience,' Kant suggests that what is *real* and objectively valid for scientific cognition is not what is merely *given* in its rough, unformed giveness, but that whose intelligibility is first *produced* and *constructed* by reason out of an independent giveness. History, we could argue, is also such a construct; it is a rational way of accounting for what is perceived as real in particular experiences. A paradigmatic example for the idea of *Erkennen* as poietic activity and rational production is provided, according to Kant, precisely by the history and *progress* of sciences such as logic, mathematics and physics.[9]

This view of scientific cognition may appear as a the reformulation of Gianbattista Vico's principle '*verum est factum*.'[10] In the *Principi di una Scienza Nova*, this principle does not apply to the construction of the physical world of nature and the natural sciences, but to the world of history and social and political institutions. For Vico, the principle *verum est factum* grounds the privileged epistemological status of historical cognition over our cognition of nature. Since history and the institutions of civil society have been *made* by man (as their *ratio essendi*), we can be confident that the principles for their cognition will be originally found in the 'modifications of our human mind' (as their *ratio cognoscendi*).[11]

However, for Kant (as opposed to Hegel), what is 'real' is not coextensive with what is produced by scientific experience. The origin of Kant's dualism between thinking and reality is the contention that human cognition is always and necessarily discursive, that is, ultimately depends upon a reality that can only be given to and by sensibility. Against a long-standing philosophical tradition, Kant radically separates sensibility and understanding, intuition and concept as the two original and heterogeneous sources of cognition. While through 'sensibility' objects are '*given*' to us, through the 'understanding' objects are '*thought*.'[12] These two 'branches' or functions of human knowledge are different in kind (not just in degree, as the tradition wants them) so that the possibility of a transition from the one to the other is utterly impossible. Because of the fundamental heterogeneity that divides them, they produce a 'synthesis' by working together. The discursive nature of our understanding, that is, its need to be complemented or synthetized with sensibility, requires us to move from the particular given in intuition to the universal of the concept. The universal does not itself directly *determine* the manifold particulars which are *given* in intuition and in sensation. The particular remains contingent in relation to the universal under which it must be subsumed, so that the particular cannot be deduced or analytically derived from the universal. Scientific knowledge is produced exclusively by the subsumptions of 'determinant judgment.' If our understanding were intuitive, then it could go from the universal to the particular. The conceptual universal would directly determine, that is, *produce* the particular, which would consequently lose its contingency.[13] The latter model is the one followed by Hegel's idea of the *Begriff*. While in Kant's model the historical reality of the universal remains particular and contingent and logically separated from the universal, in Hegel's view the universal coincides with the full development of its historical actuality: the concept is 'concrete universal.'

In order for an object to be *thinkable*, what is required, according to Kant, is simply its '*logical* possibility' which is guaranteed by the principle of contradiction. But in order to *know* an object, its '*real* possibility' should be demonstrated – either empirically by experience of its reality or *a priori* by reason. In other words, if I can 'think whatever I want in so far as I do not contradict myself,'[14] the mere thought of something does not allow me any inference about the objective reality (or the existence) of what I think. The transcendental inquiry of the *Critique of Pure Reason* rests on the delimitation, within the realm of objects, between what can respectively be known and be thought. Accordingly, all that can be known can also be thought, but not all that can be thought can be known. Counter to metaphysics, which has always mistaken the possibility of thinking things for the possibility of knowing them, the *Critique* establishes that the sphere of what can be thought is not coextensive with the realm of the knowable. The domain of pure thought is, in fact, far more extended than that of science's pure *a priori* knowledge. It includes speculative knowledge of reason, practical knowledge and the teleology of reason. Thus, it is on the distinction between thinking and knowing that Kant ultimately grounds the peculiar relation between different 'uses' of reason, and hence the chance that reason may be deceived and fall into error. On the other hand, the distinction, within scientific knowledge, between the possibility of ascertaining *a priori* by reason the 'reality' of what is merely thought (the *universal* laws of the understanding), and the necessity of recurring to empirical experience (the *particular* empirical laws of natural science) draws another fundamental separation between two orders of the 'transcendental.' The latter distinction, which grounds the later inquiry of the *Critique of Judgment*, opens up to the possibility of a constitutive role played by historical change in reality – both in the reality known by science and in the reality constructed by it.

From this presentation of the problematic of the first *Critique* we can draw the following conclusions. Kant sees the poietic nature of scientific thought as responsible for the historical 'progress' of science (logic, mathematics, physics) and for its scientific character (as opposed to metaphysics) which cannot be infringed by error or deviations from the secure path of science. The claim that scientific experience is construction of the world of nature *in general* as the inter-connected order of *universal* laws is based upon the claim that human cognition is inescapably discursive. This claim draws the limits of scientific cognition and its objective reality and validity, that is, the limits within which knowledge and reality can be said to respond to the same process and to fall under the same historical conditions. Yet the realm of particular empirical phenomena and their possible regularities and order is not constructed by the transcendental principles of the understanding. It so radically exceeds the understanding's order as to pose, for Kant, a problem that remains unresolved until the third *Critique*. In 1790 the reflective faculty of judgment will be in charge of accounting for a dimension of scientific investigation in which the empirical side of reality – the particular instance always given *a posteriori* – requires judgment to search for new laws and principles. It is in this heuristic and searching dimension of scientific inquiry that historical change (both in reality and in science's principles) can eventually assume, for Kant, its legitimate place. At the level of the reflective faculty of judgment, historical particularity receives its transcendental justification.

To be sure, in the first *Critique*, the historical progress of scientific thinking (and, in analogy to it, the alleged 'progress' of metaphysics) amounts to the discrete transition from the condition of non-science to the 'secure path of science.' This is clearly testified by the situation of sciences such as logic, mathematics and physics.[15] Herein historical progress seems to regard only the progress *toward* science, not the progress *within* science. History is reason's conquest of the 'path' to science.

An analogous idea plays a crucial role in Hegel's development of the phenomenological movement of consciousness in the 1807 *Phenomenology of Spirit*. Herein history arises at the intersection between error and truth, between consciousness's experience and repeated failure, and the pure element of the concept that grounds the *Logic* as the true and definitively assessed dimension of science.

1.2 Hegel's Speculative View and the Actuality of Science

Starting at least from the year 1800, Hegel has been thinking of his own philosophical project in terms of the 'system' of philosophy. Given his early political and historical interests, Hegel faces very soon the problem of the systematic placement of history within the whole of philosophy. Already in the Jena period, the question assumes for him a twofold form. On the one hand, in the years 1803–1806 Hegel uses the term 'Geist' to designate the collective unity of a people (*Volk*) and its role as agent of history (this is the case even when Hegel speaks, in this period, of 'absolute spirit'). In its activity, spirit is seen at work as 'alienated' in the objective world of the state and political institutions. The 'spirit of the world' works for universal history which, in turn, develops within the ethical sphere of *Sittlichkeit*. History is the dimension in which the singularity of a people attains universal significance through its action. Science as well as philosophy, art and religion are intellectual expressions of the same 'practical' spirit at work in history. To this extent, they are also inherently historical (but historical only in a subsidiary or derivative way for their 'historicity' hinges upon the historical activity of spirit).

On the other hand, since history expresses for Hegel the peculiar life of spirit, and spirit is essentially consciousness, history receives in consciousness its most proper foundation. History is therefore 'history of consciousness' (*Geschichte des Bewußtseins*). Moreover, since philosophy itself is spirit's highest form of activity, namely self-reflection and self-cognition, 'history of consciousness' turns out to be 'history of philosophy.'[16] This important link is established soon in Hegel's philosophical development. During his 1805/1806 semester in Jena, Hegel lectures for the very first time on the history of philosophy. From now on, he will consider this topic as an essential introduction to speculative philosophy.

Both lines of thought – the one that sees history as belonging to the ethical development of spirit and hence to its 'practical' dimension, and the one that stresses instead the 'theoretical' import of spirit's own activity and connects it to the development of philosophy as science – are closely linked, in the crucial years 1805–1807, to the project of a 'phenomenology of spirit.' In this period, the 'logic' which governs the internal articulation of history is a 'phenomenological' logic guided by the process of consciousness' self-cognition and coming-to-itself through alienation in its otherness. This process is viewed from the very outset as a movement of liberation of consciousness, realization of freedom, and foundation of the peculiar dimension of science. At the end of the 1807 *Phenomenology of Spirit*, the history of spirit – developed throughout the complete collection of figures of its ethical, political, and religious life – is taken up in the conclusive moment of 'absolute knowing' (*absolutes Wissen*) in which the form of scientific cognition in its pure logical dimension is first disclosed. However, the final identification of spirit's history with the 'recollection' (*Er-Innerung*) of philosophical thought that identifies the speculative dimension of *Wissenschaft*, immediately meets with a problem. The tension between the temporal succession of history and the alleged eternity of philosophical thinking leads Hegel to a further reflection on the relation between history and time. The cryptic conclusion of the *Phenomenology* seems to testify of the unresolved tension between Hegel's two different systematic interpretations of history at this time: philosophy of history or history of philosophy?

Hegel's chief concern in the *Phenomenology* is to show what the only possible form of philosophical knowledge should be, given the assumption that philosophy ought to be 'science,' and hence ought to be developed in the form of the 'system.' Philosophy is discursive knowledge *par excellence*. Moreover, it is in the dimension of reason that science has to prove both the power of its actuality and the concrete meaning of its mediations. 'The power of spirit is only as great as its expression and exteriorization [*Äußerung*], its depth only as deep as it dares to spread out and lose itself in its exposition.'[17] This is the starting point for the concluding 'recollection' attempted by absolute knowing. It is important to stress the significance of Hegel's careful remark – repeated both in the Preface and in the *absolutes Wissen* chapter – according to which the 'concept' constitutes the form or the medium of the 'existence' (*Existenz/Dasein*) of science.

In contrast to Kant's transcendental question, for Hegel is not enough to show how science is possible; what is necessary is to further display its full actuality. This need is addressed by Hegel at the very end of the *Phenomenology* by showing the necessary relation between the 'concept' of science on the one hand, and time and history on the other. Science and its 'absolute' modality of knowing are intrinsically and necessarily historical. For *history* is the actuality (and the actualization) of reason exactly in the same way in which *science* is the actualization of reason's most proper structures. In this regard, if brought back to reason as their unifying basis, history and science display homologous functions. This claim constitutes the specific phenomenological component of Hegel's project in 1807. Reason is not the re-naming of an esoteric Absolute, but the standpoint of *absolutes Wissen* or science which is reached by finite consciousness as the conclusion of the entire process of experience. Absolute knowing, for Hegel (as opposed to Kant), not only belongs to 'experience'; it constitutes its immanent structure as well as the condition of its truth. The dialectical paradox is that *truth can be said eternal precisely because it has already proved itself to be intrinsically historical*. Hence 'philosophy of history' and 'history of philosophy' are established as terms that are neither irreconcilable nor alternative. With this argument, Hegel proves both the inherent historicity of the *science* constructed by reason, and the inherent historicity of the *reality* in which reason is manifested.

This argument is taken up and fully developed in Hegel's later encyclopedic system. The result of the *Phenomenology* is to define the dimension of 'objective thinking' (*objektives Gedanke*), that is, a form of thinking in which all possible opposition of consciousness (all figures of the dualism subject–object) has been eventually *aufgehoben*. As Hegel argues at the beginning of the *Encyclopedia*, the expression 'objective thinking' means the same as – and is homologous to – the thesis that 'understanding, reason is in the world.'[18] At the level of the last moment of objective spirit, Hegel reformulates this thesis by saying that 'reason is in history'[19] as its immanent force, and hence that history is essentially a rational process. 'Objective thinking' institutes the dimension of science. Its conquest is the end result of the *Phenomenology*; its immanent development is the movement of the *Science of Logic*. Herein Hegel presents the dialectical unfolding of the pure structures of scientific thinking which designate, at the same time, the internal structures of reality. Despite its atemporal development in the *Logic*, historicity is still the mark of 'objective thinking' on the ground of its phenomenological genesis in *absolutes Wissen*.

The philosophical program of the *Phenomenology* (the claims of the Preface read along with the conclusion of 'absolute knowing') sets up from the very beginning to think the two dimensions of history – namely what can be called its practical and its theoretical dimensions – in their unity. A key move that guides Hegel in this endeavor is the final shift from 'spirit' (*Geist*) to the 'concept' (*Begriff*). The paradox of the *Phenomenology* consists precisely in its opening up, as conclusion of consciousness' own experience, to the seemingly alternative dimensions of time – that is, history on the one hand – and the concept, that is, logic on the other. Accordingly, at this level, the notion of a *begriffene Geschichte* expresses Hegel's dialectical attempt to think of the unity of time and logic, alienation and freedom, nature and spirit. This unity constitutes the dimension of science – *absolute* knowing as inherently *historical* knowing. The important point here is that, since this unity is a *dialectical* one, it is the unity of two processes (and not of two static terms). The issue disclosed by the dimension of 'absolute knowing,' that is, by the dimension of the 'system,' regards the conciliation between the historical succession in time of the figures of spirit and the eternal and timeless succession of the logical forms of the concept.

2 Science, Philosophy and the 'History of Pure Reason'

With Hegel's dialectical solution of the contradiction between concept and history in mind, we can turn now, by way of conclusion, to Kant's final chapter of the first *Critique*, where the tension between the *a priori* transcendental structures of reason and the historical and factual existence of its accomplishments in the sciences can be assessed.

The very last chapter of the Transcendental Doctrine of Method of the *Critique of Pure Reason* is dedicated to the 'History (*Geschichte*) of Pure

Reason.' To be sure, Kant immediately makes clear that 'this title stands here only in order to indicate one remaining division of the system, which future workers must complete.'[20] The history of pure reason is still an empty place in the system. Since the 'history of human reason'[21] is the history of its successes in the sciences, the 'critique' accomplishes for metaphysics the same 'historical' revolution that long ago brought logic, mathematics, and physics to the secure path of science and truth. The transcendental analysis into the structures of the human cognitive faculty which, at the beginning of the *Critique*, seemed to point to invariant functions of rationality, shows, at the end, the need to recuperate the dimension of 'history' – a history of '*pure* reason,' or a history assessed from a 'transcendental standpoint.' Even more pointedly, the title of a 'history of pure reason' is itself historical as it hints to the *future* task that the end of the *Critique* sets to philosophy itself once the systematic form of science has finally been gained. In the *Critique*, *Geschichte* is (only) the name of an absence to be filled, or (still) the beginning of a task to be carried out.

We can suggest that, in philosophy, rationality is historical *either before* the *Critique of Pure Reason*, in which case it is a history of errors and illusions, *or after* the *Critique of Pure Reason*, in which case future history becomes the task of a systematic rationality. With regard to philosophy, as opposed to the natural sciences, the transcendental philosopher, looking back in time, sees only 'ruins.'[22] History is the archeological construction of a new edifice out of the ancient ruins that the critique has contributed to destroy. Its sources are in the 'archives of human reason.'[23] Through these images, Kant presents pure reason as an edifice whose structures appear intrinsically historical.

In the light of the 'critique,' the history of pure reason is able to discern the different ages of human reason as well as the successive stages in the development of its subjective conditions. It seems to hear Vico echoing in Kant's words: pure reason's 'infancy is *dogmatic*'; then reason grows *skeptical* – wiser and more circumspect in its judgment; eventually, the age of *criticism* marks the maturity of human reason, which develops the capacity of discerning its necessary and constitutive limits beyond the empirical – and historical – limitation of its 'present' accomplishments.[24] Thus, in Kant's view, the *transcendental history* of pure reason, that is, the historicity of the transcendental structures of human reason or reason in its potentialities as subjective *Vermögen*, still remains separated from the *empirical history* of its actual accomplishments and results, that is, from the objective actuality and actualization of those potentialities. The task to which the 'history of pure reason' hints to, yet leaving it explicitly unfulfilled, is precisely a possible reconciliation between these two forms or modalities of history. The problem is structurally analogous to the one faced by Hegel at the end of the *Phenomenology*.

With regard to the present issue, however, the situation of philosophy differs radically, for Kant, from the situation of the pure sciences that do not require, in their development, the contribution of empirical inquiry. The paradigmatic case is provided by mathematics:

> If I abstract from all the content of knowledge, objectively regarded, then all knowledge, subjectively regarded, is either historical or rational [*entweder historisch order rational*]. Historical knowledge [*historische Erkenntnis*] is *cognitio ex datis*, rational knowledge is *cognitio ex principiis*. However a mode of knowledge may originally be given, it is still, in relation to the individual who possesses it, simply historical if he knows only so much of it as has been given to him from outside …, whether through immediate experience or narration, or … through instruction.[25]

It is apparent that Kant's use of the term *historisch* as opposed to rational is very far, in this passage, from the modern sense of 'historical.' Following the German scholastic tradition, historical knowledge is, for Kant, empirical *a posteriori* knowledge of given facts – *cognitio ex datis*.[26] Knowledge presents this character if taken in the 'subjective' perspective, that is, with regard to the relation between the individual and the body of knowledge that is 'given' to him and thereby 'learned'; it does not concern (objectively) the content of cognition itself. To exemplify his distinction, Kant discusses the case of the historical knowledge of Wollfian philosophy. When knowledge is merely historical, reason is at work only as an 'imitative [*nachbildende*] faculty' not as a 'productive [*erzeugende*]' faculty. The crucial point is that, in mere imitation, knowledge does not 'arise *out* of reason [*nicht aus Vernunft*]' as its true source and principle, and hence is not true *Vernunfterkenntnis*. The source of knowledge, in this case, is not reason itself but the independent giveness of historical facts, not an active reason but its products as they have become, in Hegel's language, a fixed and dead 'positive.' The one whose reason merely imitates or who knows only by way of historical knowledge is, as Kant puts it, 'a plaster-cast of a living man.' Accordingly, only the creative power of cognition out of reason, that is, of a cognition that not only objectively but also subjectively goes directly to the true source, allows one to be a truly 'living man' endorsed with a truly living reason. In other words, if it is separated from the transcendental origin of knowledge, empirical history is dead history, that is, mere imitation of a living reason. Hence, the source of living history can only be placed in reason itself, that is, *a priori*.

Kant distinguishes two kinds of *Vernunfterkenntnis*. Cognition out of reason 'is derived either from concepts or from the construction of concepts. The former is philosophical, the latter mathematical.' In the case of mathematics, the distinction between historical and rational cognition, which was drawn in the case of philosophy, does not hold. Kant remarks that it is indeed '*noteworthy* [*sonderbar*] that mathematical knowledge, in its subjective character, and precisely as it has been learned, can also be regarded as knowledge arising out of reason.' In the case of mathematics, imitation and creation (the act of learning a theorem and the creative act of the mathematician who formulates it for the very first time), with regard to the form, are one and the same act; herein the merely historical fact and the *a priori* principle coincide. In the case of philosophy, on the contrary, to learn philosophy and to philosophize are not identical activities. This lies in the fact that, in mathematical knowledge, the use of reason is '*in concreto* only, although likewise *a priori*, namely, *in intuition which is pure, and which precisely on that account is infallible, excluding all illusion and error*.'[27]

In the case of philosophical knowledge, which cannot count on pure intuition to construct its concepts, error derives from the natural destiny of reason to be subject to dialectical illusion. Herein, because of the constitution of human reason, the act of going to the source (cognition 'out of reason') may be deflected toward an illusory and false principle. This possibility of a 'transcendental' error is the ground for philosophy's need for a 'critique.' As Kant reveals in the Preface and further argues for with regard to the 'ages' of human reason, this need is an intrinsically historical need. Critique may reduce to 'ruins' all the preceding attempts to establish philosophy in scientific form, but is also the only force that can eventually secure to philosophy the path of 'science.' The idea of a (future) 'history [*Geschichte*] of pure reason' immediately follows from this argument.

Kant concludes his distinction of the two forms of *Vernunfterkenntnis*, the mathematical and the philosophical, as follows: 'mathematics … alone of all the sciences (*a priori*) arising from reason, can be learned; philosophy can never be learned, save only in historical fashion; as regards what concerns reason, we can at most learn to *philosophize*.' Contrary to mathematics, philosophy, in its very 'idea,' does not have true historical existence or actuality as a science yet (hence we cannot *learn* philosophy). Philosophy 'is a mere idea of a possible science which nowhere exists *in concreto*, but to which, by many different paths, we endeavor to approximate, *until* the one true path, overgrown by the products of sensibility, *has at last been discovered*, and the image, hitherto so abortive, has achieved likeness to the archetype, so far as this is granted to man. *Till then* we cannot learn philosophy.'[28] While it holds true that we can have only *a posteriori historical* (*historisch*) knowledge of philosophy and that the actuality of the idea of philosophy, that is, its *Geschichte* can only be thought of *after* the critique, whereby it is projected in a future yet to come, mathematics yields always rational knowledge and has indeed historical existence as science. In the case of mathematics, the path of science has been secured once and for all at the very beginning. In this path, the empirical, historical existence coincides with its rational objective actuality. While for mathematics error is a merely subjective mishap that does not affect the secure advancement of science's development (that is, the very idea of science), for philosophy (metaphysics) the possibility of error is responsible for making the beginning of its history as a science possible only after the critique.

The further consequence of this distinction, which affects the different historical status of reason's transcendental structures, is Kant's view of the relation between history and the solution of problems in science and philosophy.[29] On this issue, Kant's position differs radically from Hegel's. Science's solution of problems amounts, for Kant, to a definitive and conclusive result that sets scientific truth outside the realm of history. Once a solution has been reached, no further questioning of the initial problem can reasonably be pursued (this attempt would be a falling back into error). The 'problematic' character of reason's ideas, among which the very idea of philosophy stands, shows instead different implications. While in science 'problems' are 'demonstrable propositions,'[30] the 'problematic' nature of ideas refers to the constitutive impossibility of all solution. While with regard to scientific problems history is excluded once the definitive solution is attained, in the case of the problematic character of speculative ideas neither a solution nor an historical progress towards the solution is possible.

Hegel's rejection of the 'problematic' and merely 'regulative' status that Kant attributes to reason's ideas amounts to a conversion of the fundamental lack of reality of the 'problematic' into a reality that is intrinsically 'historical' in its process. Accordingly, for Hegel, ideas are nothing but the actual historical process through which reason's problems are formulated, solved, and yet always taken up again in varying historical constellations according to new meanings. The crucial point that guarantees, for Hegel, both the possibility of attaining a positive solution of a given problem (against its deferring *ad infinitum*) and the possibility of always reopening the problem, lies in the dialectic between the intrinsic historicity of reason's activity and the historical process of reality in which reason's realization takes place.

Notes

1 With regard to Kant, I have discussed this problem in *Das Verhältnis von Logik und Zeit bei Kant und Cassirer* (Nuzzo, 1996).
2 See Nuzzo (2002).
3 See, for example, Hegel (1955); see the *Entwurf* 1822–28, pp. 1–22.
4 See Hegel (1986), Vol. 4, pp. 64, 410.
5 For a discussion of these passages and of the general question, see Nuzzo (1989/90) and Renthe-Fink (1964); the thesis is confirned by Bauer (1963); see also Grimm and Grimm (1860); Gadamer (1908) and the article by Brunner, Konze and Kosellek (1975).
6 For a general discussion, see Margolis (1993).
7 In the vast literature on the topic, to be considered is first of all the neo-Kantian discussion around Kant's epistemology; see the fundamental essay by Ebbinghaus (1968). In the recent literature, see Friedman (1989); for the general topic and for further bibliographical information, see also Friedman (1992).
8 At the beginning of §V of the Introduction to the *Critique of Judgment* Kant clearly states the difference between a metaphysical and a transcendental principle in the following way: 'A transcendental principle is one by which we represent the universal *a priori* condition under which alone things can become objects of our cognition in general. On the other hand, a principle is called metaphysical if it represents the *a priori* condition under which alone objects, whose concept must be given empirically, can be further determined *a priori*. Thus, the principle of our cognition of bodies as substances, and as changeable substances, is transcendental if it says that change in bodies must have a cause; but it is a metaphysical principle if it says that their change must have an *external* cause' (my translation).
9 KrV (*Critique of Pure Reason*) B VIII–XIII (I quote from the English translation by Norman Kemp Smith, London: Macmillan, 1929).
10 See Baum (1984).
11 *Principi di una Scienza Nova* (1744), §331: 'questo mondo civile certamente è stato fatto dagli uomini, onde se ne possono, perché se ne debbono, ritruovare I principi dentro le modificazioni della nostra medesima mente umana.'
12 KrV B 29/A 15; cf. B 74/A 50 and the 'two original sources of the *Gemüt.*'
13 For this crucial passage, see KU (*Critique of Judgment*) §77.
14 KrV B XXVI, Fn. See this important distinction at work, for example, in the conclusions of the transcedental deduction: B 165.
15 See the already mentioned example of logic, mathematics and physics in KrV B VIIIff.
16 Fichte had already alluded to his philosophy or 'doctrine of science' as a 'pragmatic history of human spirit.'
17 *Phänomenologie des Geistes*, in Hegel (1986), Vol. 3, p. 18.
18 Enz. §24A, 25.
19 Enz. §549A.
20 KrV B 880/A 852.
21 KrV B X.
22 KrV B 880, 863/A 852, 835.
23 KrV B 732/A 704.
24 KrR B 789/A 761.
25 KrV B 864/A 836.
26 For the opposition between *cognitio ex datis* and *cognitio ex principiis* typical of German rationalism see Wolff (1740), §§1–20 for the distinction between mathematical,

philosophical and historical knowledge, §§50–54 for the historical knowledge of philosophy); see also Meier (1752), p. 157.
27 KrV B 865/A 837 (my emphasis).
28 KrV B 838/A 866 (my emphasis).
29 I want to thank Tom Rockmore for pointing our to me this further aspect of the problem.
30 See *Logik*, §38.

References

Bauer, G. (1963), *'Geschichtlichkeit.' Wege und Irrwege eines Begriffs*, Berlin and New York: DeGruyter.

Baum, M. (1984), 'Erkennen und Machen in der "Kritik der reinen Vernunft"', in B. Tuschling (ed.), *Probleme der 'Kritik der reinen Vernunft'*, Berlin: De Gruyter, pp. 161–77.

Brunner, O., Konze, W. and Kosellek, R. (eds) (1975), 'Geschichte, Historie', in *Geschichtliche Grundbegriffe. Historisches Lexikon zur politisch-sozialen Sprache in Deutschland*, Vol. 2, Stuttgart: Klett-Cotta, pp. 593–717.

Ebbinghaus, J. (1968), 'Kantinterpretation und Kantkritik', in *Vorträge, Aufsätze und Reden*, Darmstadt: WB.

Friedman, M. (1989), 'Kant on Laws of Nature and the Foundations of Newtonian Science', in G. Funke and T. Seebohm (eds), *Proceedings of the Sixth International Kant Congress*, Washington, DC: The University Press of America, pp. 97–107.

Friedman, M. (1992), *Kant and the Exact Sciences*, Cambridge, MA: Harvard University Press.

Gadamer, H.G. (1908), 'Das Wort "Geschichte" und seine Zusammensetzungen', dissertation, Freiburg.

Grimm, J. and Grimm, W. (1860), 'Geschichtlichkeit/Geschichtlich', in *Deutsches Wörterbuch*, Leipzig: Verlag von S. Hirzel.

Hegel, G.W.F. (1955), *Die Vernunft in der Geschichte*, ed. J. Hoffmeister, Hamburg: Meiner.

Hegel, G.W.F. (1986), *Werke*, Vols 3 and 4, ed. E. Moldenhauer and H.M. Michel, Frankfurt am Main: Suhrkamp, 1986.

Nuzzo, Angelica (1989/90), 'Storia della filosofia tra logica ed eticità: considerazioni sul ruolo e la collocazione sistematica della 'idea' di filosofia in Hegel', *Annali dell'Istituto Italiano per gli Studi Storici*, pp. 259–301.

Nuzzo, Angelica (1996), 'Das Verhältnis von Logik und Zeit bei Kant und Cassirer', in M. Plümacher and V. Schürmann (eds), *Einheit des Geistes. Probleme ihrer Grundlegung in der Philosophie Ernst Cassirers*, Frankfurt am Main: Lang, pp. 59–80.

Nuzzo, Angelica (2002), Hegel's Method for a History of Philosophy. The Berlin Introductions to the Lectures on the History of Philosophy (1819–1831)'. in D. Duquette (ed.), *Hegel's History of Philosophy: New Interpretations*, New York: SUNY Press, pp. 19–34.

Margolis, J. (1993), *The Flux of History and the Flux of Science*, Berkeley, CA: University of California Press.

Meier, G.F. (1752), *Auszug aus der Vernunftlehre*, Halle: J.J. Gebauer.

Principi di una Scienza Nova (1744), Napoli ([1953] ed. F. Nicolini, Milan: Ricciardi).

Renthe-Fink, L. von (1964), *Geschichtlichkeit. Ihr terminologischer und begrifflicher Ursprung bei Hegel, Haym, Dilthey und York*, Göttingen: Vandenhoeck and Ruprecht.

Wolff, C. (1740) *Philosophia rationalis sive logica*, Francofurti and Lipsiae, Officina libraria Rengeriana.

Chapter 4

Historicity, Social Psychology and Change

Rom Harré and Fathali M. Moghaddam

1 The Very Concept of Historicity

The dictionary definition of 'historicity' is 'historical quality or authenticity based on fact.' Historicity is a quality of a discourse perhaps, or even of an opinion. The most general concept would something like this: the historicity of a claim about the past is its factual status. The multi-faceted discipline of psychology implicitly makes claims about the past (and the future) by its resolutely ahistorical stance. The results of contemporary studies are presented as, and probably believed by mainstream psychologists to be, universal, pan-temporal laws of human nature. 'Mainstream,' official psychology of human beings is atemporal, or to put this another way, it assumes that the organizing principles of human association have always been the same. This presupposition is surely contestable. We shall contest it by critically examining the practice of contemporary social psychology. To make the force of what we have to say clear, a preliminary sketch of social psychology is in order.

Social psychology is typically and (vaguely) defined as the scientific study of social behavior.[1] Even a cursory survey of the literature makes it clear that neither term in the *definiens* is uncontestable. Neither a grasp of the nature of the scientific tradition nor a clear idea of social behavior is evident in this literature. This issue has been worked over many times. Our interest is in something rather different, the role of temporal assumptions in the practice of psychology. Social psychology, as such, first appeared in the United States about one hundred years ago in the writings of two major figures, a psychologist, McDougal,[2] and a sociologist, Ross.[3] Although social psychology is typically American in its emphasis on individuals as the sites of the relevant psychological processes and its dependence on experimental methodologies, it is a rapidly growing specialty around the world.[4] The expansion is in effect due to the export of doctrines of American social psychology reflecting American culture. This material is marked by its atemporality. The issue of the historicity of social psychological claims, reports and doctrines in the contemporary context sketched above, is clearly of the first importance.

The issue of the historicity of the discipline is also of importance because of the relative closeness of it to history proper. Some scholars have argued that social psychology can be regarded as history,[5] while others have gone so far as to declare that it *is* history.[6] Such questions as 'What is the "historicity" of the hypotheses of social psychology?,' 'What is the status of assumptions about the psychology of people of the past?,' 'How do these compare with assumptions about the chemistry of the geological past, the biological assumptions given form in alleged facts of descent described in the genealogy of a family?,' and so on require examination.

Even though there are very few social psychological studies that look back to past historical periods, in several important ways social psychology has been 'historical.' For example, undergraduates reading social psychology books come across many studies that were published before they were born, such as those by Sherif[7] and Milgram.[8]

Gergen has argued that in reading social psychology of another era, students are reading social history.[9] Social psychology just is the keeping of a record in past terms of past aspects of past forms of social life. Now comes the problem of how these documents can be read in the present. How do students from this era interpret the social behavior of students (in laboratory experiments) in earlier eras?

There seems to be a much greater problem with the historicity of claims about what people believe, think, feel and so on, rather than what they do. The historicity of a description of the Battle of Borodino, for instance, as to the number of casualties is quite different from the historicity of Tolstoy's famous remarks about Katusov and his attitude to the battle. Yet even the casualty figures are subject to the problem of historicity. For example, the figures given by Shakespeare for the Battle of Agincourt (hundreds of French dead, a couple of dozen English) look 'massaged'!

But if there has been change in meanings, conventions and so on between then and now, how would one know it? How would we get a handle on what psychic life was like before the end of the Napoleonic era if our psychological lives are different? The hermeneutic circle seems to close very tightly around this problem.

This investigation clearly involved several matters:

1. By what criteria or method do we sort out traces and remnants of the past in some particular order? Need the chosen order be the simple order of modern clocks or calendars? The choice of the order of clock-time presupposes some sort of mapping from cause/effect sequences to significant sequences of events as time sequenced. Thinking in terms of intentional actions of people with projects on hand might lead to highlighting the significance of events in a different hierarchy. Only if causes are what matter does a narrative have to take the picaresque format.
2. How we know how to categorize an event as a *social* event? How do we demarcate or individuate such events? Are there pan-temporal criteria for arriving at the very material on which a historically sensitive social psychology could be based?
3. Items 1 and 2 together drive us towards the idea of a systematic process or processes of *social* change. But is this any more than a myth? What would be needed to ground the idea of social *change*, particularly if that idea were to be grounded in the idea of the changing psychology of the people who live a different kind of social life? Sexual arrangements have greatly changed in the last fifty years. What is the quality of the claim that people in the 1950s were different in some psychologically significant way from the people of the twenty-first century? This is to question the historicity of theses derived from alleged examples of social change.

We cannot discuss the historicity of a claim, a putative factual claim about 'what happened,' unless we have taken some sort of stand on the three issues above.

How would one determine the historicity of a claim about the psychological processes of Genghis Kahn, Adolf Hitler or Albert Einstein? So to discus 'historicity' in the social psychological/psychological content thoroughly, one would need to examine the broad claims of psycho-history. In this chapter we will concentrate our attentions on the field of social psychology because the historicity of its claims is most fundamental and has been most strongly contested.

2 The 'Historicity Problem' in the Context of Social Psychology

2.1 The Absence of Time in Social Psychology

At first glance, it would seem that traditional social psychology is not seriously concerned with time, with the past. At most there is a paradoxically atemporal focus on the future. About 85 per cent of social psychological studies concern brief episodes, typically about an hour, involving undergraduates in laboratories. The main objective of such studies, often explicitly stated, is to identify causal factors leading to specific behaviors, in order to achieve more accurate predictions about future behavior. But the methodology ensures that future behavior must be the same type as behavior that can be identified in the hour or two in the 'laboratory.' The small number of field studies conducted by social psychologists also adopt this 'causal' objective. For example, a central goal of social psychologists over the last century has been the prediction of behavior from declarations of attitudes.[10] The idea that wholly new forms of behavior might emerge from certain identifiable conditions *cannot* be realized within this 'experimental' methodology.

2.2 Neglect of the Past

A major reason for a lack of interest in the past is the research design accepted in traditional social psychology. This design calls for the manipulation of independent variables (assumed causes) to measure their effect on dependent variables (assumed effects), holding all other factors constant. Given that it is not possible to go back in time and manipulate variables in past social life, it is not surprising that traditional social psychology focuses on (slices of) the present, in order to arrive at predictions for future behavior.

A small number of social psychologists have attempted to study the past in order to better understand social behavior. McClelland's original studies on achievement motivation come to mind.[11] The main proposition underlying these studies is that economic profess arises from certain attitudes and behaviors that add up to a 'need for achievement' (an idea not unlike the Weberian notion of 'Protestant Ethic'). To test this proposition, McClelland developed measures of a 'need for achievement' and reviewed different historical eras, from the time of the Roman Empire to the twentieth century.[12] He claimed that each era of economic prosperity was predicted by a rise in the 'need for achievement' as measured by his instruments.

A 'finding' from a psychology experiment represents a point in time, a particular juncture in process. We show our biases by identifying one point in a process as a 'finding,' while other points are neglected as 'non-findings.'

3 The Very Idea of Change Presupposes Continuity

There could be no perceived change unless there were perceived stabilities and continuities.[13] This cliché sounds like a profound metaphysical principle. But on reflection it begins to seem much more like a linguistic rule: no event sequence should be called 'a change' unless there is also opportunity for the application of phrases like 'remained the same.' However, this change of status does not diminish the central importance of the principle in everything to do with human life.

There is another principle, less frequently cited and so less of a truism. It is that history is narrative. A catalog of events lacking any of the characteristics of a story is empty, even for such dry matters as the geology of one of the moons of Jupiter. Without an implicit principle of progression, development, teleological thrust or something similar, there is no story and so no history.

Reflections on the stabilities and instabilities of social life must be constrained by these two principles above all. Yet there is something deeper to be seen in both. The perception on temporality and the phenomenon of memory are inseparably bound up. This observation too may seem to be hackneyed. What could be more obvious than that for a change to be perceived some recollection of a relevant previous state must have occurred? Research into the apprehension of melody, an ideal test frame for hypotheses about temporality and cognition, has shown that when studied closely, the hackneyed observation above fragments into a variety of memorial processes.[14] Summing up the results of a great deal of work we can say that changes are perceived only with respect to some implicitly recollected frame. For instance, whether the interval C to F is a change from tonic to subdominant or from dominant to tonic depends on whether the key that has been established is C major or F major. The auditory experience of the same frequency ratio is phenomenologically different in the two frames. Not only does one have to recollect the events of the past in considering the status of an event in the present, but one must maintain throughout a certain frame of reference within which both past and present take on a certain stability of meaning.

In what follows we shall be criss-crossing a textual landscape, in a pale imitation of the techniques advocated by Witttgenstein, as we try to give an account of historicity as it is manifested in the temporality of *discourses* in the realm of the social. The Leitmotif will be an emphasis on presentations of continuity and conservation, on the persistence of themes and practices, against which we experience social life as a flux. Our emphasis on continuity is in line with recent writings by historians who point out the perils of 'periodization': the tendency to categorize and apply stereotypes to people living in different historical periods. Periodization has been associated with a tendency to exaggerate differences between such loosely defined epochs as 'The Middle Ages,' 'The Dark Ages,' 'The Roman Period' and so on. This leads to a neglect of important continuities.[15] A succession of disconnected and discontinuous happenings is not a pattern of social change. It is, if and when it occurs, a frightening descent into chaos. Paradoxically, the more closely we are concerned with change, that which at first sight seems to be the essence of temporality, the more we will emphasize continuity.

A point of central importance for the historicity of a claim is whether we see change or continuity depends in part on the magnification of our analytical lens. Using low magnification, we look at long stretches of time and tend to see discontinuities and change, while with high magnification, we look at short stretches and tend to see continuities. A good example of the different views obtained at different powers of magnification is the alleged scientific revolutions of the seventeeth and the twentieth centuries. From the hegemony of a Catholic Aristotelian conception of the universe in the fifteenth century to a Protestant Newtonian picture in the eighteenth century, a huge change, there intervenes three centuries of gradual shifts and refinements. This is one of the sources of the 'periodism' we drew attention to above. At a sufficiently low magnification, there do seem to be distinctive periods, the boundaries between which seem to vanish when the magnification is increased.

4 Bearers or Carriers of Continuity

If we are right about the 'logical' structure of the concept of change, namely that it must always be understood in tandem with the concept of continuity, what are the bearers of social and historical continuities? What is it that is the same at a later time? And what is it that changes, so that it is possible to perceive that time as later? So to apply the concept of change, there must be ways of applying the concept of continuity. From a positivistic standpoint, all one would need to do would be to compute practices, institutions and so on at different times to see whether indeed there were sufficient and relevant similarities to declare that some had persisted unchanged. But the deeper question is this: what lies behind the practices whose comparison one with another we can make by simply observing them? Here we must ask about the kinds of bearers of continuity that there might be. The principle involved here is that practices are grounded in rules or rule-like habits so that the variety of such normative constraints provides the basic taxonomy of bearers of continuity. How do we upgrade fragmentary relics of the past into a full-blown historical fact? By what discursive devices is the transformation achieved? It is not by digging deeper into the subsoil of Anatolia or finding more dusty documents in the library of the Escorial! Identifying something as a 'bearer of continuity' *establishes* something as the basis for perceptions or judgments of historical continuity. It is not a discovery.

4.1 A Typology of Carriers or Bearers

Still using the biology analogy, just as bodies are interpreted by Dawkins as vehicles or 'carriers' for genes,[16] we can conceptualize carriers for norms rules, and other aspects of normative systems. Such carriers are of enormous variety, but we can discuss them in a more manageable way by classifying them as public or personal, opportunistic or stable.[17]

Let us begin by considering simplified examples of each type. The stars and stripes is a public carrier, in the sense that this flag represents of Americans the values, traditions and so on of their nation. Each new generation of children in US schools learns to chant the allegiance to the flag, and this public act is seen to be a continuous thread linking children with past generations.

But a particular flag might serve as a personal carrier. Joe always has on his person a tiny flag given to him by his friend Jack, who died in Vietnam. Joe does not show this flag to anyone else, but each time he brings it out of his pocket it reminds him of the promises he made to help Jack's wife and children. The flag embodies these promises, as well as his resolve to keep his word. One day Joe gets fed up with the disrespect shown to him by Jack's family, and in disgust throws away the tiny flag, deciding that he has done all he can to honor his promise to the dead friend. Nobody else ever learns about the flag and what it has meant to Joe, but the throwing away of the flag is of great significance to him. It represents a break with the past, and a dramatic change in his behavior.

The public/personal distinction blurs in most real-life cases, because the same carrier has both a public and personal role. This is particularly the case when other people serve as personal carriers, in the sense that in the public domain they represent certain values, ideals and so on, but they also have special personal significance as carriers in a private way for each individual. Part of the mystery of 'fame' and 'charisma' is explained by this: famous people like the late Princess Diana have an enormously wide appeal because they successfully act as carriers in both personal and public domains. She both represented the traditions and values associated with some aspects of British aristocracy, and she became integrated as a carrier in the private lives of countless individuals, she had personal meaning and symbolism. Roger Brown's notion of 'flashbulb memory' captures some of this idea: John Kennedy's assassination was a public event that served to act as a 'flashbulb' for private memories, everyone of that generation could remember where they were and what they were doing when they heard the news of the assassination.[18]

Above all, carriers are plastic and malleable. Most carriers are opportunistic, in the sense that they are adopted for a particular purpose and will be abandoned if they fail to serve the required function. There has been in the 1990s an ongoing battle between supporters and opponents of the 'old' flag of the State of Georgia. The former view the flag as 'carrying' important and cherished traditions of the Old South, including Southern hospitality and chivalry. The latter see the flag as carrying hated traditions, such as slavery and inequality. During interviews with flag supporters, participants said they would 'fight' for the flag, but they often also added that if the flag was changed they would adopt other symbols to represent the Old South. Of course, such changes involve costs. One cannot invest in and abandon carriers without costs.

Novelists rather than psychologists have best captured the personal suffering people experience when a carrier important to them fails to perform adequately. A dramatic example is provided by George Eliot in *Adam Bede*, the story of a talented young carpenter who, like everyone else in the region, considers Arthur Donnithorne, the heir to the great aristocratic estate in the region, to represent progress as well as valued traditions. But Adam makes a terrible discovery: Arthur Donnithorne has seduced a young girl much below him in rank and status, a girl he could not possibly marry. In that instant, the gallant carrier of great aristocratic traditions is transformed to an unfeeling boy only concerned with satisfying himself. But because of the carrier role this boy had, much more is transformed:

> For the rest of his life he [Adam Bede] remembered that moment … as a man remembers his last glimpse of home where his youth was passed, before the road turned and he saw it no more.[19]

The historicity of a huge range of such past-tense statements that is inherited from the status ascribed to carriers plays a major role in such cultural matters as identity, national or ethnic.

Carriers, then, are also 'investments,' in the sense that people use them as anchors for their emotional and social worlds. Giving up a carrier can be costly, and may involve giving up aspects of the world that are valuable to us. Carrier reference or carrier talk established the historicity of a discourse for all practical purposes. 'This was the *very robe* my grandmother was Christened in,' 'I was named 'Emmeline' after her!' and so on.

4.2 Public Carriers across Historical Eras

A remarkable feature of life in many parts of the world is continuities across long periods, hundreds and sometimes thousands of years. How is such continuity achieved? Part of the answer to this question is revealed when we consider the role of certain public, formal and fairly stable carriers. Religious ceremonies, such as marriage and christening, come to mind. But we shall take our example from a non-Western culture, to consider the role of *Ta'ziyeh* in Shi'i Islam.

Ta'ziyeh is a Shi'i Muslim tradition about the martyrdom of Hossein, the third Imam of Shi'i Muslims and the grandson of the Prophet Mohammed. Hossein was killed in 680 AD, in the Islamic lunar month of Moharram. Hossein and his followers are said to have suffered greatly, being without food and water for many days before being cut down on the plains of Karbala, in present-day Iraq. *Ta'ziyeh* is performed throughout Iran, as well as other regions where Shi'i Muslim communities are found. *Ta'ziyeh* preformances are enormously popular, particularly among the poorer sections of the population.

Ta'ziyeh may appear to Western audiences to be something like a mixture of opera and drama, with the 'good' characters chanting their parts in melodic verse, but the 'bad' characters portraying their evil nature by the way they say their lines. *Ta'ziyeh* performances can be extravagant and lengthy; one of us attended a three-hour performance in Iran where some characters came 'on stage' on horseback and whole armies of supporting players were involved. The success of the performance is judged in large part by how much it moves the audience from one emotional state to another, and particularly on how much the audience is moved to weep. In this respect *Ta'ziyeh* is similar to another very important ritual in Shi'i Islam, *rauzeh-khani*, which typically involves a man (the *rauzeh-khan*) reciting the story of Hossein and moving an audience to weep for the martyrs of Karbala. What *Ta'ziyeh* achieves through action drama, *rauzeh-khani* tries to achieve through power of language alone. Obviously the latter is more economical.

Although *Ta'ziyeh* enjoys popular support among the masses, several powerful groups have in the past attempted to co-opt it to suit their own purposes, and even to ban it. *Ta'ziyeh* has not enjoyed support from some conservative mullas, in part because it seems to go against Islamic rules forbidding making images of The Prophet. Also, *Ta'ziyeh* attracts attention and money away from the clergy, and puts it in the hands of actors, artists and other such 'unholy' types. *Ta'ziyeh* has had a love–hate relationship with the Shahs of Iran, being formally banned for several decades in the first half of the twentieth century, but revived in the 1960s and 1970s.

A number of important values central to Shi'i Muslim culture are carried by the *Ta'ziyeh*. A first is the acceptance of ones fate, even if that involves defeat and death. But, second, this is not necessarily a fatalistic value system, because the *Ta'ziyeh* can also be interpreted as a call to arms, a determination to fight *even in the face of enormous disadvantages*. Shi'i Islam is a 'minority religion,' not only because of the break away from the main Sunni Muslim Body, but also because just about every Shi'i leader has died at the hands of enemies.

5 A Metaphysical Assumption of this Way of Posing the Problem

But aren't we already slipping into the assumption that the past is there to be accessed if only we could do it well? The very idea of a carrier as a methodological concept presupposes that we can recognize a past carrier or past version of a carrier as such. But of course, the past is not there in the required sense. It is not like something to be dug up, a common enough metaphor for the past-directed enterprises. A lot of common ways of talking, replicated in social scientific discourse, might make it might look as if the foundation of the factual quality of a claim about a past event in the social world is just a special case of the correspondence principle of truth. But even if records survive, they exist only in the present, so correspondence between present claim and 'past' fact is no more than just another case of hypothesis juxtaposed to hypothesis. This point has been made before, but we need to use it to address the status of social psychological 'facts' in particular and hypotheses about 'them.' Our thesis is quite simple: such facts exist only as items in narratives.

Historical narratives are about people, nations and so on, pinned together more often than not by the continuity of the being about whom the narrative is told. Autobiographies illustrate how it can be that a story of shifts and changes of fortune can yet be the story of a person, one and only one human being. The central features of the sense of temporality in human affairs, the tensions between change and continuity, neither of which would be intelligible without the other, is modeled rather well in the telling of stories about oneself.

5.1 Narrative in General

Narratological studies of many discourse genres, including the reporting of scientific experiments,[20] the presentation of environmentalist arguments and so on, have revealed the somewhat surprising fact that the overall structure of most tellings and writings exemplifies story-telling conventions rather than patterns of logically ordered premises and conclusions or straight forward chronicles of a 'this happened then that happened' sort. The power to persuade for example seems to reside more in the plausibility of a traditional story line than it does in the fulfilling of the strict criteria of logic. One of the most powerful analytical devices for revealing the step by step pattern of a discourse is the scheme of Vladimir Propp, originally abstracted from a study of the plots of folk tales.[21] Propp identified more than thirty steps that appear in the plots of traditional stories. Every folk tale draws on some, and always in the order they lie in the ideal totalized plot. Thus the hero suffers a loss, is sent on a quest, receives the help of a powerful being, and eventually triumphs. The same patterns are evident in Greek myths and the Homeric tales. The stories we tell about or own lives and the lives of others, the fictions we elaborate to instruct or amuse, follow a limited number of patterns, patterns the narratologists call 'story-lines.' Not surprisingly, one cluster of such patterns can be seen to fit into the Proppian repertoire. As Harriet Hawkins has argued, classics and trash obey the same dramatic conventions.[22] *Othello* and *Neighbours* share patterns of character relations and of plot. This must surely be a matter of interest to psychologists intent on explaining the unfolding of episodes of human interaction, and in the longer run, the development of a life among the developing lives of others.

The strongest narratological thesis that would link the study of the patterns of story-telling to the problem of explaining human action is this: 'Lives are lived according to the same conventions in accordance with which lives are told.'[23]

This thesis has profound consequences for the concept of autobiography. It is the fact that an autobiography has both retrospective and prospective dimensions that makes autobiographical telling a prime subject for psychological research. No only do we tell ourselves and others versions of the lives we have led, but we tell ourselves and others anticipatory stories that express the pattern of those parts of our lives that are yet to be lived. Shakespeare was very good at presenting this aspect of human psychology.

His grasp of this phenomenon appears in such famous soliloquies as that of Richard III over the corpse of his predecessor and that of Hamlet attempting to resolve the existential dilemma at the heart of his struggles to find a way to deal with the murder of his father. Should his autobiography end in suicide or in revenge?

Anticipations of this line of thought can be found in Bruner's recent discussion of autobiography,[24] and in the contributions to the understanding of murder by Jean-Pierre de Waele.[25]

5.2 Autobiographies

At first glance, the concept of autobiography as a narrative genre could hardly seem more innocent of complications. It is just the story of my life as told by me, from my point of view. Autobiographies may differ in degree of candor and self absorption, but who is the best authority on what I did and why I did it than myself? But reflection on the matter coupled with empirical studies of autobiographical narration quickly discloses all sorts of complexities.

An autobiography is above all a narrative, and in each age narrative genres have their own conventions. Caesar's *Gallic Wars* and St Augustine's *Confessions* share some narrative conventions but not others. Both are self-exculpatory. But Caesar wrote *The Gallic Wars* as the plain tale of a bluff soldier (however disingenuous that style may appear to us).

He begins his 'simple tale' thus: 'Gallium in tres partes divisum est,' as boring a geographical observation as one is likely ever to encounter. However, the great 'one-liner,' 'vidi, veni, vici' is more appropriate to the implicit drama of the narrative. St Augustine's psychological epic is as much a story of the 'inwardness' of a soul at odds with itself as it is of public action. It could hardly be more different from the style of Caesar's fragment of autobiography.

But an autobiography is a narrative in the first person. Grammatical studies show the first person is a complex indexical device expressing at least four aspects of the user's sense of self, as a singularity in arrays of other people, ordered by various and shifting sets of relations, spatial, temporal, moral and social.

An autobiography is not just one story, my story, told to some generic and anonymous listener or reader. The written tale is just one of the autobiographies that a person did tell or that they could have constructed. Research into autobiographical narration in everyday life discloses how the quality, value, detail and arrangement of the episodes recounted depends on the person to whom the tale is told, the context of the telling and the aim of the story-teller at that moment in the telling of it.[26] Everyone has a multiplicity of potential autobiographies, though few may see the light of day. That multiplicity is evident in the practice of retrospective telling. But, as I pointed out above, at least as important in the everyday practice of autobiography is the prospective telling, how my life is going to evolve, in the next hour, the next day, in the next decade, and in eternity. Autobiography not only reports and interprets action, it shapes action.

It follows directly that behind pathological lives there must lie pathological story-tellings. So it is to autobiography, as the most psychologically relevant form of narrative for the understanding of disturbed and distorted lives, that we must turn. Of course, the very idea of 'disturbed' and 'distorted' presupposes the existence of patterns of lives that are 'correct' or 'normal,' and that must surely be a topic for the anthropologist and historian to explore. At any moment in the history of human psychology there will be patterns that are the taken-for-granted background of uncontentious ordinariness. But what was ordinary for the Toltec (see Pizzaro's *The Conquest of Mexico*) or a medieval Carmelite (see *Heloise and Abelard*) might seem very strange and even pathological to a St Albans shopkeeper in the 1990s.

Who is the subject of an autobiography? One's sense of self, though elusive, is certainly among the experiences that are ineluctably private. But what sort of experience is that? Not, it seems, of some inner entity. While we try to work with the assumption that talking about ourselves is descriptive and that 'I' refers to something, we will be inclined to wonder to what should our personal history be ascribed. However, if we drop that assumption, encouraged by Wittgenstein's general distinction between describing and expressing, we are close to a resolution of the seeming paradox that the phenomenological elusiveness of 'the self' threw up. If words like 'I' are used to express structural properties of experience, and to present ourselves as responsible for what we say and do, there is no 'thing' to which 'I' refers and which is being described when we express how we feel or what we are thinking. The only singularity around which anyone's life is built is the person they are. Once we have accepted that much of speaking is expressive, we can turn to ask what linguistic acts give public expression to the sense of self. A fourth element in a person's sense of self is the sense of living a life in time. The pattern of events in which one has a sense of the uniqueness of one's life is a multiply ordered pattern of various stories each of which recounts retrospectively recollected and prospectively anticipated events. There are many ways of telling one's life, each of which has some claim to our allegiance. In different circumstances and for different purposes and to different audiences, different stories are told. And yet each of us, as the author of a multitude of personal narratives, had little difficulty in maintaining them all as versions of our one and only life.

Unitas multiplex is a motto which captures the most characteristic feature of our own lives, and is effortlessly achieved by most of us.

One's brain and nervous system provide one with recollections, but language provides one with memories, and memory provides one with a host of ordered series of recollections, fragments of autobiography. One's autobiographies provide one with a life. Language and other symbolic devices permit one to anticipate the future, by imaging events in which one could have a role. For most people these futures too are multiple. It is language that provides one with the possibility of autobiographies, recollections indexed not only by the indexical 'I,' but as located in sequences of recollections ordered as stories by reference their indexing as past, present and future with respect to their temporal relation to acts of telling. This grammar certainly facilitates and perhaps makes it seem natural to suppose that there is just the one *timeless* person persisting through the various versions of that one person's life events that are remembered as that cluster of stories we are tempted to call *the* autobiography. It seems that it is the indexicality of the first-person singular that allows many versions seamlessly to express one life.

The temporality of the sense of self, the singularity of a continuous trajectory in time, is not a simple function of the ability to present just one autobiography, since most people not only can but do have different stories to tell about themselves. Studies of how people tell their lives show that each of us, having a sense of our singularity in the three dimensions of our relations to other embodied persons, experience our life retrospectively and prospectively in relation to more than one autobiography. Like positions, autobiographies are functions of the situation and persons to whom they are told, including autobiographical soliloquies, which are more often than not directed to some imagined other.

Furthermore, as we have pointed out, autobiography has both a backward and forward direction. There is not only what one is telling oneself about one's past, but also what one is telling oneself about one's future. The future dimension of autobiographical telling is no more unitary than is the past. At each moment one locates one's present self at a moment, this moment, on some suitable world line, and so establishes one's temporal singularity as a self. But that trajectory may be abandoned for another at some subsequent moment. Whereas in space one has and can only have one self, since one exists in one and only one body, in time one can and does have many selves. How is this possible? Provided we keep the two major senses of 'self' distinct, there is no paradox. Self as singularity is different from a set of beliefs about the person one is. While in relation to the body the former must be singular in non-pathological life stories, the latter is under no such constraints. Whereas the criteria of identity of human bodies are such that at one place in the material world there can exist only one body at a time, say at the moments of an act of speaking, the criteria of identity for events, as they form elements in the temporal trajectory of a life, are potentially multiple. There is always the interpretation that lies between sentences and the statements they are used to make, between actions and the acts they are used to perform. Indeed, the study of real symbolic interaction shows that most linguistic and other symbolic acts are to some degree indeterminate.[27] Sometimes the situation requires that we must make them more precise, but we rarely need to bother. There are huge numbers of events in our lives, so the possibility of making different selections for different purposes also makes multiplicity possible.

5.3 Pronouns as Indexicals

In telling a tale in the first person, one is committed, everything else being equal, to the four indexing forces of the use of the pronoun 'I' and equivalent grammatical devices. Thus a report of what has been seen, heard or touched by me is indexed with the place of my body, the time of my speaking, the position I occupy in the local moral order, and in some cases, with my social position too. All of these indexing acts can be qualified. An alibi will place the embodied me far from the scene reported, and the uses of tenses, in tales told in an Indo-European language, will modify the temporal relation of my act of speaking to what that speaking describes, commits me to and so on. Autobiographical telling in ordinary life involves the claiming and disclaiming of responsibility for actions, for what has occurred. The ordinary indexical force of the first person is to take or claim responsibility positioning oneself as an agent. This is, as linguists say, the unmarked use. Unless it is explicitly repudiated, the use of the first person is agentive. Therefore, it indexes what has been done with the moral standing of the actor, and so with the level of responsibility that position entails. To see how this works, we must lay out a sketch of the grammar of agentive discourse.

Usually when 'me' is preferred to 'I,' the implication is that the actions reported are not those of the speaker as agent. Roughly, 'I did it' and 'it happened to me' are the poles of a simple contrast between speech acts in which the speaker presents him or herself as agent or as patient. When I appear as patient in a fragment of autobiography, I am repudiating responsibility for what has happened. Things which 'just occur to me' are not brought about by me.

However, this contrast is too crude for much that must be achieved in autobiographical discourse. To be an agent is to have the power or capacity to do something, if unimpeded. The generative explanation of what happens takes its start from an act of the person responsible for the action. If the situation is right, my powers and capacities, tendencies and intentions, projects and plans will be executed if there is no impediment to the activity or power or lusts of the person involved.

When we bring this aspect of agentive discourse into line with the way in which the normal indexical force of 'I' expresses my sense of responsibility for an action, a further level must be examined. There is a contrast between 'I do what I intend unless I am prevented from doing so' and 'I indulge my lust for Belgian chocolates unless I am prevented from doing so'; they are at opposite ends of the spectrum of discursive devices for taking and repudiating responsibility. In the former example, the implication is that I arrived at an intention by a process of reflection that I myself initiated, and that the impediments are independent of my wants and powers, while in the latter case the implication seems to be that the lust for these famous and delicious confections is not something I brought about by reflections (and hence not my responsibility), while there is at least the weak implication that the impediments placed in the way of such self-indulgence might be self-erected.

Studies of fragments of autobiography offered by convicted and indeed self-confessed murderers[28] brought to light a discursive convention that routinely presented the speaker as patient. One might make a case, from the frequency and taken-for-granted character of the use of this convention that 'telling a personal story in the patient mode' is the unmarked version. The phrase that carries implication of passivity is 'and then *I caught a charge ...*': The model for this construction is something like 'I caught a cold.' Things you catch are personal states and conditions all right, but they are, as it were, out there, floating about, and by chance and through no fault of your own you run into them. A killing may be reported as something that happened, in a neutral style that neither takes nor repudiates responsibility, but that the police should hold me responsible for it, to the extent of charging and ultimately trying and even condemning me, is something that position the speaker in their own moral order as 'the one who should be held responsible for doing it.'

The fourth component of the indexical grammar of English story-telling is the indexing of what has been described or avowed as an event in the life history of the speaker. Lacking the tensed first-person pronoun, English speakers must use tense to order their autobiographies, and at the same time, the continuity of the story-line is ensured by the transtemporal sense of 'I' and 'me.' The same is not true of 'we,' since the collective, membership of which it expresses, can and does change with time. Lying about one's past and fantasizing about one's future are obvious temporal pathologies of the autobiography. But in this paper we are concerned with pathologies which infect the indexicality of the story-telling. It is not only pronouns that bear temporal indexicalities, but also tenses. What if someone lacked a sense of the past and had no anticipations of the future? The first person might survive in their autobiographical tellings, but there would be no verbs inflected for tense. Or perhaps some reportings would be indexed as past and others as future, but with those categories no order would be expressed, because, according to the discursive point of view, no order in recollection would be experienced. In the case notes of psychiatrists (for instance, in those of the redoubtable Oliver Sachs, in whose consulting room some remarkable cases seem to congregate) there must be material that would be of the greatest interest to those discursive psychologists who are interested in the indexical pathologies of autobiography. There is lots to do in charting these divergences and using their light to reveal what, in this culture or that, counts as the right and proper away to order a life.

It might seem obvious that an autobiography is a window into its author's soul. But pathological 'souls' can find expression either in the unusual content of their stories or in the use of a strange grammar. The study of expression of self in stories is part of discursive psychology. This development is based on a Vygotskian thesis about the shaping of mind in the learning of linguistic and practical skills in symbiosis with another person, and on a Wittgensteinian insight that how we feel and how our thoughts are organized are expressed in characteristic language-games. The selfhood of autobiographical telling is expressed predominantly in the uses of first- and second-person (indexical) pronouns and in the choice of narrative conventions within which to tell the story. Pronouns are used to index what is said with the various locations of the speaker as a person among persons in several patterns of relations.

In summary, we can see how the uses of 'I,' together with the tenses of verbs, and in accordance with local narrative conventions, express the shapes of the many stories we can tell about ourselves. It is quite usual for each person to have many autobiographies. But when non-standard uses of pronouns appear, we must be alert to a kind of linguistic pathology which, given the strength of the expressive account of language uses, may express a pathological structuring of the mind of the speaker.

6 The Nature of Time

One thing that would knock a great hole in any attempts to use correspondence theory of truth in historical social psychology would be a demonstration that only the present exists, and nothing but the present has ever existed. The concepts of past and future are narratological, not existential,

There are two arguments towards the conclusion that there is only the present.

6.1 The Grammar of Temporal Discourses

The Character of 'Now' One might be tempted to think of the word 'now' as a special kind of name or referring expression that picks out a certain moment in the flow of time, within which the events of life are embedded. There are all the nows, the river of time. And then there are the events which match some of them. However, 'now' is an indexical. It indexes the temporal relations of an event to the event of the utterance of a statement describing or prescribing and so on that event. Thus 'now' is a relational expression between two events. It is not the name of the moment in time. There are no such moments. There is no frame of pure moments of time. To think there is, is just to confuse the grammar of indexicals with that of nominative expressions. There are no past 'nows', nor are there future 'nows' to come. Some past events have been simultaneous with the utterance of descriptions of them, and some future events no doubt will be similarly related to the utterance of descriptions of them. But neither past nor future exist as temporal regions, strings of moments, to be contingently occupied by events.

Narratives as Conjunctions and as Disjunctions There is another 'grammatical' source of the illusion that as well as events there is a time manifold for them to occur in or with. The illusion can be seen clearly if we highlight the difference between the logical structure of a historical *narrative*, and the logical structure of the totality of existence claims that support or are presupposed by the narrative.

A historical narrative is a conjunction of statements which purport to be true. A happened *and* B happened *and* C happened *and* so on. The logical form is as familiar to logicians as that of truth function '&,' the properties of which include the principle that one false conjunct renders the whole false. If, however, we examine the corresponding sequence of statement asserting the existence of the events in question, then the logical form is disjunctive. A exists *or* B exists *or* C exists and so on. This is the truth function 'vel,' and obeys the principle that so long as there is just one true clause, the whole disjunctive statement is true. This is just what we would want for a grammar of time as the sequence of events which actually exist.

There is no temptation to create a sempiternal manifold out of the disjunctive form. However, since each clause of the conjunctive form must be true if the whole story is to be true, there is a temptation to think that somehow there is a queer kind of way in which the past, present and future coexist, as referents of each of the true statements comprising the narrative. The grammar of conjunction tends to case a shadow on the world. It might appear as the metaphor of time as a river, of life as a journey or some other of the models that tend to draw the imagination in to creating a picture of a events as a kind of panorama.

Seeing the matter in this light helps to strengthen the point about 'now.' There is no manifold of nows either, a dimensionless sequence of pure temporal moments within which the events of the everyday would occur.

The point has been nicely made by Latour.[29] We have all this surviving stuff, records, artefacts, human memories and so on, left-overs from all sorts of ears and 'times.' The attic is full of this rubbish. We go about sorting the items according to some principle or other. Maybe we arrange them so that what is most important comes first, tailing off into the irrelevant. Or we use some conception of a pattern of cause/effect sequences which some philosophers have suggested is the foundation of our sense of time. Thus historicity, the very idea that there is history as a record of past facts, and that we have lived it and are still living it, is only one of the many possible organizing principle for currently managing what does not now exist, whether because it already has been or because it has yet to come to be. 'It is the sorting that makes the times, not the times that make the sorting.'[30]

6.2 *Narratives through which Social Time is Created*

Tradition Another candidate for the bearer or carrier of continuities in social order, and thus the forging of the historicity of the discourse, is tradition or traditions. At first sight it is above all the traditions of a community, its immemorial traditions, that define it for what it is. We think of a traditional Christmas, traditional weddings and so on. This appears to be at least the revival or preservation of forms from a former time. The very notion of 'Christmas' has that quality of the past in the present we call 'historicity.'

But there is more to it. If asked 'Why do we have turkey at Christmas?' we would not be surprised to get the answer 'It's a tradition.' But in a way that is not an answer at all. We suggest that it is a distinguishing mark of traditions that they do not admit of anything but a pleonastic answer to the question 'Why do we do that?' In some ways this is like 'It's customary' as a way of stopping a regress of why questions. But not the same. There is in the case of tradition a sense of antiquity, of an origin lost in the mists of time, certainly before living memory. And that has some of the power of imposing the authority of the tradition.

Lévi-Strauss remarks that there may be nothing special about the event that initiates a tradition.[31] It must, however, occur at some appropriate moment, and be distinguished enough to be readily and naturally repeatable. For example, at Linacre College, Oxford, there was insufficient money to provide a tray for everyone having lunch. This meant that people had to return their trays after they had taken their meal to the table. So Linacre people do not eat off trays. Wolfson, well endowed to begin with, had trays for everyone, so that there people do eat off their trays and return tray and dishes when they are done. However, Linacre people will from time to time express their sense of superiority in the quality of life at their college by remarking that at Wolfson people are so uncouth as to eat off their trays.

Of course, this points to the other aspects of tradition, that traditions serve to characterize and to maintain the character of a particular community. Traditions such as the Catholic meatless Friday, seriously under consideration for reinstatement, serve to maintain a sense of belonging to a particular Christian sect, and perhaps to do more, reminding people of the ascetic side of Christianity, of the legend that the crucifixion was on a Friday and so on.

Custom If one is advised on a procedure by being told 'It's the custom here,' say to hang one's coat on the banisters, the implication seems to be that it is what is generally or as a matter of course done. There is little implication of the sanction of antiquity, nor that something else is achieved by adhering to this pattern, as the use of the word 'convention' would imply. The traditional anthropological distinction between 'law' and 'custom' fits in with this account, since a custom is something that is done routinely, by habit, rather than in conformity with an instruction or an edict. Custom may be pan-temporal, but is not, we believe, a source of historicity.

Heroic Exemplars The celebration of President's Day and Martin Luther King Jr Day come close together in the calendar of national festivals in the United States. On these days Washington, Lincoln and King are mentioned, their achievements are described, excerpts from their speeches are repeated and so on. From a more objective historical viewpoint, none of these great personages can compare with Thomas Jefferson as the architect of the United States, both literally and figuratively. Yet he has no 'day' in which his virtues and his influence are celebrated. The rise of Lincoln as a repository of national virtues, that is, as a carrier or bearer of a sense of national continuity, is instructive, since the sense of historical past to which his life refers is a retrospective construct-historicity of a discourse as the result of the projection of the narrative forms of the present and some typical devices into the 'grammars' of discourses marked as tales of the past.

As Barry Schwartz says:

> Every society ... displays, and perhaps even requires, a minimal sense of continuity with the past and its memories cannot be serviceable to the present unless it secures this continuity with the past and failed to outlive changes in society, the society's unity and continuity would diminish ... Society's memory of its great men is one part of this 'symbolic code' [through which a sense of national identity is maintained].[32]

It was not until the early years of the twentieth century that Lincoln became 'the most cherished of national possessions.' Schwartz argues that it was largely through the extraordinary national celebrations of the centenary of Lincoln's birth that his status was transformed. But why Lincoln, and why that time? Schwartz points out that until that time the identity of the United States turned on the myth of origin, in the heroic figures of the National Congress at Philadelphia, where such men as Washington, Jefferson and Hamilton created a nation in their own image. After all, Jefferson wrote the words, designed the buildings, and bought most of the real estate. But this was an America of an educated elite. The new century required the mythic celebration of equality, the American Dream, new minted, of the man from nowhere who rises to lead the nation in the very conflict in which the ideal of equality was embedded. The shift in focus was most visible in a new genre of writings that compared Washington and Lincoln, in which the fame of the latter began to outstrip that of the former, as the emphasis shifted toward a nation seeing itself as a democracy and distancing itself from its past. The creation of exemplars creates a continuity that did not exist before them. Continuity is to a large extent retrospective. National memories are not true recollections of the past, but creations through which a national past is forged.

7 Models for the Processes of Social Change

It seems that historicity is a property of texts and discourses. It is a way of managing the structure of a discourse, so that it develops according to one possible sequencing mode, namely that of the order of temporal existence of the events in question. We might then be tempted to think that causality was the deep operating principle behind the up-front temporality, supporting historicity as the reflection of causality. Thus the historicity of a claim would be established by working causality 'backwards' so to speak, retrodicting to *what must have happened* and what caused it to happen. However, such a simple solution to the ordering problem posed by Latour is easily seen to be unsatisfactory. There are many difficulties in trying to set up a scheme for understanding social change in which the causal processes are presumed to be simple, and positive. A past event or state is singled out, *ceteris paribus*, from a background of stable states, as the cause or part of the cause of whatever succeeds it. This did not work in biology, nor does it work well in the social sciences. It seems that selectionist explanations, negative causality, what exists in what is left over from causal processes of the deletion of those alternatives not fitted in some way to survive are unacceptable.

Philosophers of science, for example Toulmin[33] and van Parijs[34] have noticed that there are two rather different formats for the construction of historical explanations. Some formats make use of the idea of positive causality, that some state or condition or event that exists or occurs at a certain time produces, generates or is a sufficient condition for the occurrence of some subsequent state or event. The past or some aspect of the past brings it about that some states of affairs comes to be in its future. The complexity of interaction factors, such as the project developing a positive explanation on anything but the smallest of scales, would be worth pursuing.

However, there is another format, that of negative causality. The conditions of the past destroy or frustrate the originators of all options but the one which survives. Darwinian evolution theory is the most famous historical explanation that is couched in the format of negative causality. But at the level of generality at which that theory would serve as a model for an account of social change, there is insufficient detail to provide convincing accounts of change. Remember that in the scheme of negative causality, change is a kind of illusion. The overall picture has changed because the alternative possibilities that might have co-existed with the survivor have been eliminated. But the survivor has not changed.

We can look to a further refinement of the biological analogy to flesh out an account of temporality in the social world as negatively causal. Since it is the genetic material that is conserved rather than the bodies of the organisms that realize it, the focus of negative causality must be in that material, or to put it roughly, on genes. Generalizing the distinction, biologists have used the terms 'replicator' for the gene or gene-like entity, and 'interactor' for the body or realization of the gene in an environment, in which its fate will determine which genes survive into the next generation. Changing patterns of gene distribution look different, and it is they, but not the genes, which have changed. To complete the model, one must then ask what it might be that could serve the replicator function in social matters, and what the realization of such entities as interactors with the environment might be. Dawkins called them 'memes,' but we shall find that the concept of rule, taken as including both implicit rules and explicit rules or instructions, will cover most cases. Interactors will be the practices that realize the rules.

The negative causality of selectionist format for social matters has the following layout:

1 Rules are taught to new members of the society if explicit, or picked up by imitation if implicit. They serve the role of replicators. Each new generation of people will have its characteristic distribution of rules. Some rules will become very widespread in a population, while others will die out.
2 People carry out the practices of social life in accordance with or by following rules. It is practices that succeed or fail in the local environment. If a practice does not succeed, then it is less likely that the rules behind it will be taught to or picked up by the next generation.
3 New rules will serve to introduce new practices.

The upshot will be that the rules that lie behind the least unsuccessful practices in the existing environment will spread through the population.

In this way meme-rule theory serves to account for social change. But there are some reservations to be noticed. In the biological model it is only the interactors that are tested and succeed or fail. But in the social model there are many occasions in which people imagine carrying out some practice and discuss whether it is appropriate or not before they try to put it into practice. Sometimes this is enough to abandon a proposed novel rule or instruction. Parliaments are institutions set up mostly for this purpose. In biological evolution the genes of one generation can reach the next one only by being passed through the intermediate generation. But rules can be revived after many generations. For instance, Hebrew, a language long dead, was deliberately revived, by teaching the rules of this linguistic practice, and these rules were passed on. Finally, we should notice that sometimes it may seem strained to interpolate a 'rule' stage between the practices of one generation and another. Could not practices be learned directly by imitation? But if we are to take this seriously as a psychological theory of change, we must suppose that the person who has learned the practice has acquired a skill and is not simply copying the master on every occasion.

8 Conclusion and Summary

Social psychologists, despite many reminders, have generally failed to address the question of the historicity of their alleged discoveries and hypotheses. Locked into the attractive idea that the human world will yield universal laws of behavior more or less similar to those yielded by studies of the material world, the very idea that there might be an issue of the historicity of textbook 'results' is rarely entertained.

However, when we open up this issue we quickly find ourselves immersed in very deep problems, just because the issues are one of the forms of discourses. Narratives are assumed to be narrations of descriptions of sequences of events. But reflecting on the grammar of temporal indexicals like 'now' and the logic forms of such narrations, we quickly see that we never effectively break out of a discursive realm. We are always and necessarily assessing stories in relation to one another. If the coherence theory of truth has a place in philosophy, it is surely in commentaries on the methodology and metaphysics of social psychology and related enterprises.

What is it in narrative that creates the sense of history, of past, present and future? We suggest that the answer is simple in outline, but complex in practice. Both continuity and change, mutually necessary to one another, are displayed in a great variety of devices all of which are carriers of the sense of the historicity of a narration. However, there are other ways of ordering the events described in historical narrations, such as relative significance, legality, character display and so on.

Notes

1 Moghaddam (1998).
2 McDougal (1908).
3 Ross (1908).
4 Carr and Schumacher (1996).
5 Gergen (1973).
6 Billig (1995).
7 Sherif (1965).
8 Milgram (1974).
9 Gergen (1973).
10 Lapier (1934).
11 McClelland (1961).
12 Ibid.
13 Moghaddam (2002).
14 Sloboda (1996), pp. 52–5.
15 See discussion in Golden and Tooley (1997).

16 Dawkins (1976).
17 Moghaddam et al. (2000).
18 Brown (1966).
19 Eliot (1859 [1961]), p. 290.
20 Nash (1990).
21 Propp (1968).
22 Hawkins (1990).
23 Bruner (1991).
24 Bruner (1993).
25 De Waele (1976).
26 Helling (1977).
27 Pierce and Cronen (1980).
28 O'Connor (1994).
29 Latour (1993).
30 Ibid., p. 76.
31 Lévi-Strauss (1966).
32 Schwartz (1990), p. 82.
33 Toulmin (1972).
34 Van Parijs (1987).

References

Billig, M. (1995), *Banal Nationalism*, Thousand Oaks: Sage.

Brown, R. (1967), *Social Psychology*, New York: Free Press.

Bruner, J.S. (1991), 'The Narrative Construction of Reality', *Critical Inquiry*, Autumn, pp. 1–21.

Bruner, J.S. (1993), 'The Autobiographical Process', in R. Folkenflik (ed.), *The Culture ofAutobiography*, Stanford, CA: Stanford University Press, pp. 38–56.

Carr, S.C. and Schumacher, J.F. (eds) (1996), *Psychology and the Developing World*, Westport, CT: Praeger.

Dawkins, R. (1976), *The Selfish Gene*, New York: Oxford University Press.

De Waele, J.P. (1976), 'The Personality of Individuals', in R. Harré (ed.), *Personality*, Oxford: Blackwell, Ch. 7.

Eliot, G. (1859 [1961]), *Adam Bede*, Boston: Houghton-Mifflin.

Gergen, K.J. (1973), 'Social Psychology as History', *Journal of Personality and Social Psychology*, 26, pp. 309–20.

Golden, M. and Tooley, P. (eds) (1997), *Inventing Ancient Cultures: Periodization and the Ancient World*, London: Routledge.

Hawkins, H. (1990), *Classics and Trash*, New York and London: Harvester/Wheatsheaf.

Helling, I. (1977), 'The Carpenters of Konstanz', in R. Harré (ed.), *Life Sentences*, Chichester: Wiley.

Lapier, R.T. (1934), 'Attitudes vs. actions', *Social Forces*, 13, pp. 230–37.

Latour, B. (1993), *We Have Never Been Modern*, trans. C. Porter, Cambridge, MA: Harvard University Press.

Lévi-Strauss, C. (1996), *The Savage Mind*, Chicago: Chicago University Press.

McClelland, D.C. (1961), *The Achieving Society*, Princeton, NJ: Van Nostrand.

McDougal, W. (1908), *Introduction to Social Psychology*, London: Methuen.

Milgram, S. (1974), *Obedience to Authority*, New York: Harper and Row.
Moghaddam, F.M. (1998), *Social Psychology: Exploring Universals across Cultures*, New York: Freeman.
Moghaddam, F.M. (2002), *The Individual and Society: A Cultural Integration*, New York: Worth.
Moghaddam, F.M., Slocum, N., Finkel, N., Mor, T. and Harré, R. (2000), 'Toward a Cultural Theory of Duties', *Culture and Psychology*, 6, 3, pp. 275–302.
Nash, C. (ed.) (1994), *Narrative in Culture*, Routledge: London and New York, Chs 1–5.
O'Connor, P.E. (1994), 'Narratives of Prisoners', doctoral dissertation, Georgetown University.
Pierce, W.B. and Cronen, V. (1980), *Action, Communication and Meaning*, New York: Praeger.
Propp, V. (1968), *The Morphology of the Folk Tale*, Austin, TX: University of Texas Press.
Ross, E.A. (1908), *Social Psychology*, New York: Macmillan.
Schwartz, B. (1987), *George Washington: The Making of an American Symbol*, New York: Free Press.
Sherif, M. (1965), *Attitudes and Attitude Change*, Philadelphia, PA: Saunders.
Sloboda, J.A. (1996), *The Musical Mind*, Oxford: Oxford University Press.
Toulmin, S. (1972), *Human Understanding*, Princeton, NJ: Princeton University Press.
Van Parijs, P. (1987), *Evolutionary Explanations in the Social Sciences*, Totowa, NJ: Rowman and Littlefield.

Chapter 5

The Reality of History

David Carr

History as a branch of knowledge begins with a distinct handicap. While there may be serious disputes about whether theology, for example, has any object, everyone agrees that the object of history does not exist at all. In view of this fact, it is perhaps no wonder that skepticism about history's claims to knowledge has always been widespread.

Reasons for this skepticism are not hard to come by. The events of the past cannot be seen, heard or felt, and any assertions we make about them must be grounded by the most indirect means. Testimony to their existence is often such that we cannot be sure even of its meaning, much less its truth. What is worse, historians may be even less trustworthy than the evidence they examine. The personal, political, religious or other prejudices of the investigator seem more likely to affect the study of past human events than they affect the study of animals, plants or inanimate nature.

In our own day skepticism abounds regarding the objectivity and truthfulness of even the physical sciences. Many feel that the 'scientific' pretentions of the so-called human sciences, history in particular, need even more obviously to be deflated. One form of skepticism about history has arisen from reflections on the narrative form in which historical knowledge is often presented. If historians are essentially telling stories about the past, their activity seems more literary than scientific. The standards of story-telling are different from those of truth-telling: the point is to produce a coherent tale with beginning, middle and end, and perhaps with a moral lesson to convey. The proper place for narrative is fiction, which is by definition unconcerned with the reality of the events it portrays. Story-telling about real events thus runs the risk of being inadvertantly fictional, more concerned with aesthetic than with scientific criteria. On this view, history seems condemned by the very form in which it is written to distort or misrepresent the events about which it claims to know.

Against this skeptical view, it can be argued that the very reality of history – *res gestae*, which are human acts and experiences, plans and projects

– already has the narrative form in which historical writing is largely cast. Far from being merely the literary form of either fictional *or* historical writing, narrative is found at the level of human events themselves. Story-telling is, to use Ruesens's expression, the manner in which natural time is transformed into human time.[1] For Paul Ricoeur, similarly, 'time becomes human to the extent that it is articulated through a narrative mode.'[2] Agreeing with Ruesen, Ricoeur and other theorists on this point, I have argued elsewhere at length for the narrative character of ordinary human experience and action, both individual and social. I have claimed that this form is found even below the level of explicit story-telling and is characteristic of the way time is humanly experienced and structured.[3]

If this theory is true, it would undercut the skeptical challenge to historical and other narratives. It could not be said that they impose an alien structure on the realities they deal with, systematically distorting them in the process. Far from differing in structure from historical reality, historical narrative shares the form of its object, and can be seen as an extension and refinement by other means of the very reality it is about. This theory thus affirms a continuity, rather than a disparity, between historical reality and historical narrative.

I am convinced that this 'continuity theory' (as I shall call it) is substantially correct. But there are some serious challenges to this theory which have not yet been met by its defenders, myself included. The theory seems to me to be particularly vulnerable at two points, and in what follows I would like to focus on these weak points, examine the corresponding challenges as I see them, and defend the theory against them.

The two points I refer to are located at the two ends of the continuum the theory affirms. The first concerns the narrative character of ordinary experience and action. Against the claim that narrative structures are found at the root of our temporal and practical experience, it can be argued that there are other ways time can be experienced and organized, and that the continuity theory in fact invokes a parochial conception of time – one that is specifically Western, and perhaps even modern – in its analysis of human experience.

If the continuity theory can respond to this challenge, it faces another one at the other end of its continuum. It may be that historical narrative is simply a sophisticated and specialized form of the same story-telling that takes place in the world of ordinary experience, and is thus continuous with it. But one also has to account for how it differs from that 'ordinary' world, and this turns out to be very complicated. It could be argued that the motives, principles and intentions of historical narrative are so different from those of non-historical narrative that their similarities seem insignificant by comparison. Most important, one must take account of the fact that historical narratives are not just stories that differ in various ways from the stories that characterize ordinary actions and experiences; they are also *about* those stories. That is, they are second-order narratives whose purpose is to tell about the ordinary (and extraordinary) actions and events of individual and especially social life. What is more, they seek to evaluate these first-order stories in a disinterested or objective way. This seems to impart to historical narrative a character that so sets it apart from the 'first-order' level of narrative structures that the idea of continuity is seriously challenged.

It is these two challenges, then, that I propose to take up in turn in what follows.

1 Narrative and the Everyday Experience of Time

It is easy enough to question the narrative analysis of everyday life on grounds that are derived from a simple reflection on one's own experience. Stories bring coherence to the flow of events, shaping them into wholes with beginnings, middles and ends, grouping them around characters with persisting identities and personalities, introducing problems, crises, turning points only to lead them toward satisfying solutions and conclusions. How different our lives really are! Incoherence, if not downright confusion, seems more typical than coherence: rather than actually beginning, things just gradually emerge; plans are interrupted, things go wrong, loose ends are left hanging, nothing is ever settled, no problems ever *really* solved or resolved. Nothing ever ends: each apparent ending is just the beginning of something else.

It is no doubt this sort of consideration that has led theorists like Louis Mink and Hayden White to affirm the artificiality of narrative. 'Stories are not lived but told,' says Mink. 'Life has no beginnings, middles and ends … Narrative qualities are transferred from art to life.'[4] According to White, 'no given set or sequence of real events' has intrinsically narrative features; they aquire these features 'only by the imposition of the structure of a given story type on the events.'[5]

Now it seems obvious to me that these theorists are operating with distinctions – life versus art, real versus 'imposed' – that are not very clearly thought out. The 'real' events of 'life' are after all not a meaningless sequence of unrelated nows, but rather the actions and sufferings of persons and groups of people. To be sure, sometimes these are chaotic and confused. But this is only because we judge them – indeed, experience them – by standards of coherence that function as the rule in our daily lives. By the term 'rule' here I do not mean merely that our lives are more often organized than disorganized, our plans and projects more often successfully carried out than not. This will differ from individual to individual and from time to time. Rather I refer to the temporal values that are part of the *sense* of events as we experience and live through them.

The point is that human events and actions derive their sense from their relation to past and future, that is, from their place in a temporal configuration in which they follow from something and lead up to something else. The idea of coherence in human affairs derives from the very way we experience and exist in time, not from the imposition on it of alien or artificial categories. Only because of this can we experience and lament the contrasting incoherence and confusion that so often plague us.

Thus the deepest and most significant structures we find in narrative – beginning/middle/end, problem (or crisis)/solution, suspension/resolution, reversal of fortune and so on – are structures that characterize our very existence as human beings. This is not just the way we *tell* about our lives after the fact; this is how we live them from the start. In this case it turns out to be true that art – at least in its narrative form – indeed imitates life. Before it is a literary genre, or indeed a form of historical writing, narrative is the practical and 'existential' form of human time. It is through the implicit construction (and constant revision) of a life-story that the individual achieves or aquires an identity.[6]

Moreover, it is not merely as individuals that we exist in this way. Groups of persons too – those we refer to as communities – can be said to constitute themselves around and through narratives. Just as the individual exists through the implicit life-story, so the community exists through a 'story' which draws together the shared memory and expectation or projection. Here too, the social present derives it sense from past and future.

Such then, briefly, is the response of the 'continuity' theory to the above objection. Narrative structure is not 'imposed' on anything. It constitutes the principle of organization for our action and experience, even though it may fail as often as it succeeds. It is the structure not merely of a certain kind of writing or even a certain kind of knowledge; it is the structure of our very being. As such, it is the very 'reality' of history. This is the sense in which the term 'historicity' has been used in the philosophical tradition, and some version of this view was held by older philosophers of history, even though they did not speak directly of narrative. Vico's notion that we can understand history because we are the makers of history suggests the analogy of form between historical existence and historical knowledge. The same is true of Dilthey's statement that 'we are historical beings first, before we are observers of history, and only because we are the former do we become the latter.'[7] Hegel believed that it was no accident that the term *Geschichte* denotes both the *res gestae* and the *historia rerum gestarum*: 'we must suppose historical narrations to have appeared contemporaneously with historical deeds and events. It is an internal principle common to both that produces them synchronously.'[8]

But this quote from Hegel's lectures on the philosophy of history already, perhaps inadvertantly, points in the direction of a deeper and much more serious challenge to the theory of continuity. Historical narrations have not always existed: they are not found in non-literate and so-called traditional societies; Herman Kulke's paper informs us that even a highly sophisticated and literate culture, like that of classical and medieval India, can be without any demonstrable sense of or interest in history.[9] Does this have any significance for the continuity theory? Not necessarily. This theory claims that historical narration derives from the narative structure of human existence, but not that this narrative structure always gives rise to historical accounts.

But this new consideration nevertheless brings out something about this theory that introduces a suspicion. Narrativity is being affirmed as a *universal* human characteristic. Claims are being made about the temporal structure of *human* experience, indeed human existence, as such. Yet some of the most important and by now 'classical' work in anthropology and comparative religions – I am thinking of C. Lévi-Strauss and M. Eliade respectively – has suggested that 'peoples without history' have a completely different way of experiencing time itself.[10] I find this conclusion being reaffirmed, if I understand him correctly, by Klaus E. Mueller in his paper 'Praehistorisches Geschichtsbewusstsein.'[11]

Is it possible that we have put forward as a universal human trait what is in fact only one way of experiencing time? In order to answer this question, it may be necessary to engage in more empirical research into the mentality of other peoples. But prior to that, we can ask ourselves if we can imagine experiencing time in any other way. According to what was said above, time is experienced in such a way that the present derives its significance from the past and the future. But what if the true significance of the present derived instead from its relation to what is divine and out of time altogether? It is important to note that we are talking about sense and significance, not just chronological order. Thus the present may differ from the past chronologically; but this difference may not be experienced as important. Sameness, rather than difference, may be the most significant relation of present to past, and indeed to future as well. Such are the elements of a non-historical conception of human time, one based on repetition, ritual and commemoration. Everyone recognizes that there are cyclical elements in our everyday experience of time – the pattern of bodily functions, the rhythm of day and night, the changing of the seasons. Nor is it the case that the cyclical is confined to the 'natural' as opposed to the cultural: the weekly calendar, the pattern of work and rest, the alternation of social and personal, public and private lives also follow a plan of repetition whose chief value is stability rather than change.

If we can recognize in our own everyday existence elements of an experience of time which is very different from the 'narrative' pattern described earlier, we may be inclined to look at the latter in a very different way. The tendency to tell stories about our actions and experiences, the linear-teleological conception of time, the interest in the past which leads us to preserve and write history about it, may all be different aspects of an approach to time which is specifically Western. Kosellek writes about the shift in European thought from prophesy to prognosis, in which the envisaged future come under human control and management.[12] Both sides of this transition belong to the same general conception of time as consisting of qualitative change moving toward the future. But the advent of Western modernity intensifies the narrative conception of time and diminishes the influence of cyclical features that persist in the largely religious ideology of the premodern period.

From this point of view the narrative interpretation of time would be seen not as the way human beings *per se* exist in and experience time, but as the way modern, Western human beings exist in and experience time. To affirm this as a universal trait would be to consign non-historical peoples to a status that is not – or perhaps not yet – fully human.[13] After contrasting the modern with the pre-modern experience of time in this way, the next step is to envisage a postmodern experience of time, in which the values of narrative coherence would be supplanted. And indeed this is suggested by many of the theorists of the postmodern. We are invited to consider the possibility that the diverse events we live through, as individuals and as groups, do not and *should* not add up to a coherent story at all, and that the belief that they do is at best wishful thinking, at worst a form of violence we do to each other and even to ourselves. The rejection of narrative goes hand in hand with the rejection of 'instrumental' reason and technology. Ultimately it leads to the rejection of history itself and of the concept of the subject, especially that of the subject of history – 'man' as hero and central character of his own story. Terms like 'collage,' 'pastiche' and 'bricolage' express values held in higher esteem than coherence and unity.

Can the continuity theory respond to this challenge? I want to propose a two-levelled response. First, even if narrativity is not a *universal* character of human experience, it is still possible to affirm the *continuity* which is at the heart of the theory, and to deny the discepancy or discontinuity alleged to exist between historical narrative and human reality. What I mean is this: even if the historical consciousness which leads to historical writing is largely Western and even modern, it is still reflective of, and thus continuous with, the reality of modern Western existence. It is still true to say that narrative structure pervades the way 'we' live our lives, rather than being imposed upon it from without. It is just that the 'we' is no longer the universal 'we' of humanity, but is admittedly and openly the 'we' of modern Western life, however far that designation may reach. If 'historicity' and 'narrativity' are features of Western, or even Western modern, rather than of all humanity, they are still features that are found at all levels of life, from everyday experience and action to the manifestations of literary culture, including historical research and writing.

But perhaps it is not necessary to give up the point that narrativity is a *universal* feature of the human experience of time. Here I come to the second level of my reponse. Above I proposed, in lieu of an empirical investigation, that we reflect on a possible non-narrative experience of time. Recall that 'we' found it quite easy to articulate what such an exprerience might be, and 'we' were able to discover elements of it in 'our' own experience. But if 'we' (Westerners? Moderns?) can find these in our experience, perhaps non-Westerners, non-moderns, even *post*moderns, can find elements of the narrative conception of time in their experience. Perhaps there are two (or even more) different ways of existing in time, or different aspects of existing in time, one of which is linear, teleological, narrative and historical, while another is cyclical, repetetive and stable. Perhaps we could say that both aspects are equally universal or equally human, but that each has been brought to the fore and manifested in cultural forms by different societies at different times. This would allow for a full vindication of even the universal claims of the continuity theory, while opening the door to an even richer and more elaborate account of the basic human experience of time.

To some extent the foregoing consists of speculations about forms of human experience that can best be filled in by concrete empirical research. But the risks and dangers of the empirical investigation of 'mentalities' are well known, and it is hard to see what 'observation' would be like when dealing with the human experience of time. Any such investigation stands in need of flexible conceptual guidelines; perhaps these reflections can serve in that capacity.

2 Historical Reality and Historical Narrative

We now turn to the second major challenge to the continuity theory. The claim of this theory is that historical narrative – the research and writing which express historical knowledge of the past – is on a continuum with historical reality, ultimately sharing the same form and constituting a 'higher-level' expression of the same features we find in everyday life. But the theory must still account for what happens in the transition from one end of the continuum to the other. The problem is: can we account for this transition in a way which does justice to the distinctness of historical knowledge while still maintaining that the continuum exists?

One way to deal with the transition in question is to focus attention on the relation between the individual and the social. When we speak of the narrative character of ordinary life as the way (or *one* way) human beings organize their experience of time, it is natural to reflect on one's own individual life, and to think of the temporal structure of one's own actions and experiences. My analysis of this structure has been largely inspired by the phenomenological accounts of temporality found in such philosophers as Husserl and Heidegger. These thinkers devote their attention primarily to the conscious life and existence of the individual, often to the neglect, according to some of their critics, of the social dimensions of human existence.

History, of course, is about social reality, and is concerned with the lives of individuals only to the extent that they figure in social life. If historians devote their attention to individuals, it is typically to important political figures or, at the opposite end of the spectrum, 'ordinary' individuals regarded as representative of some broad social or political phenomenon. If it is going to account for the connection between history and ordinary life, our theory must somehow relate individual and social aspects of time.

But as we have seen, it can be shown that the narrative organization of time is a pervasive structure of social as well as individual life. Where the first-person plural becomes as important as the first-person singular in social discourse, we can think of the cohesive community as a kind of large-scale individual which is the 'subject' of certain kinds of experiences and actions. Within such a community, persons are related to each other not merely as individuals facing other individuals, but as members of the same group. For individuals who stand in this relation of membership, experiences, actions, opinions, even feelings are not 'mine' but 'ours.' Where the communal 'we' is operative, whether at the family, ethnic, political, religious or even the universal-human level ('we' landed on the moon in 1969), social time can be said to be structured along the same narrative lines as individual time.

This is the sort of consideration which leads to the familiar notion that history is 'society's memory.' As memory is to the individual, so history is to the community. As the individual implicitly or explicitly composes a life-story which constitutes his or her identity, so the community is composing its own biography in the history it writes for itself.

The problem with this familiar notion is that it does not really do justice to the specificity of historical inquiry as we think of it today. The idea of history as a *Wissenschaft* appealing to critical standards of objectivity may be a fairly recent one, but it does exist, and the philosophy of history needs to take it seriously – though, of course, not uncritically. If the continuity theory is to do its job, it must at least find a place for historical knowledge in this modern sense.

It seems to me that the notion of history as society's memory, while extremely important, is really at a different place on our continuum. Here it is not only the narrative structure of individual experience that is replicated at the social level, but also the underlying motives that go with it. By that I mean that the community holds on to its past for the *practical* purpose of organizing and orienting itself in time, and of maintaining its identity in the face of the threat of fragmentation – a threat which is faced to a greater or lesser degree by any community. It is here that the past is kept alive in the traditions, legends, monuments, public buildings and folk music and art. No doubt popular history belongs here, too, and even the officially chosen history books of the public schools. But the underlying motivation here is practical, and it is the values of group cohesion, good citizenship and so on, which guide the maintainance of the public memory.

What goes along with this, it seems to me, is that the primary orientation of this social organization of time is, as with the individual, toward the *future*. In the broadest sense, as noted already, it is concerned with sustaining the community against the threats of an uncertain future. But it is also connected with action, with 'getting things done.' This is why it would be a mistake to think of the 'social memory' as something static. On the contrary, it is always being appealed to and subtly manipulated, especially in the context of political rhetoric. A community may owe its coherence to a generally accepted social narrative, but that does not prevent its dividing into factions, and factions often differ – for example, the progressives from the conservatives – by virtue of attitudes toward the past. But the past is important in this context because of how it affects the future. Is it a golden age from which we have fallen away? Then we must return to the values of our fathers. Is it merely the first steps toward a goal as yet unrealized? Then we must continue to move toward the future. In the political arena, the struggle is often for control of the past. But the past is prized because of its significance for the future.

In one of my earlier writings on this topic, I had occasion to cite as an example of political rhetoric, in the best sense of the word, Abraham

Lincoln's Gettysburg Address. This famous speech, known to every American schoolchild, has recently been brilliantly analysed by the classicist and historian Gary Wills.[14] Wills's detailed treatment confirms my view that Lincoln's speech is paradigmatic in a discussion of the role of narrative in the political context. A few references will show how this is so.

The President's brief remarks, delivered on 19 November 1863, were part of the ceremonies dedicating a military cemetary at the site of a fierce and crucial Civil War battle that had occured just four months earlier. Lincoln spoke for only about three minutes, but his words were telegraphed across the country and almost immediately achieved the iconic status they stll have today. According to Wills, all those who heard Lincoln's message, admiring its eloquent simplicity and directness, accepted along with it, almost without knowing it, a major reorientation of American history.

Wills locates the speech in the tradition of military-funeral oratory extending back to Pericles and Gorgias, all in the context of the Greek revival of nineteenth-century America. Lincoln's words, he says, have 'the chaste and graven quality of an Attic frieze.'[15] He notes that, like Thucydides' version of Pericles, Lincoln assumes the communal 'we.'[16] 'Four score and seven years ago *our* fathers brought forth on this continent a new nation …,' he says. 'Now *we* are engaged in a great civil war, testing whether that nation … can long endure.' What struck me, of course, though it it is not emphasized by Wills, is the narrative character of the address: the occasion, an event in the life of the nation, is part of a story, the story of the American republic itself. It derives its sense from an origin or foundation in the past, and from an envisaged future placed in jeopardy by the present crisis.

On the surface, the hortatory purpose of Lincoln's oration is obvious: 'It is for us the living … to be dedicated here to the unfinished work.' The war must be won so that the great experiment, begun 87 years before, shall not fail, so that 'government of the people, by the people, for the people, shall not perish from the earth.' Wills's analysis shows that Lincoln's rhetoric is expressive of a deeper and more radical purpose: a re-casting of the nation's past. Not the Constitution of 1787, which emphasizes confederation and states' rights, but the Declaration of Independence of 1776, is the founding document of the United States. This document, 'dedicated to the proposition that all men are created equal,' permits Lincoln an oblique reference to the slavery question. But even more important, according to Wills, Lincoln believed that in the Declaration, 'Americans had constituted themselves a single people long before the Constitution was drafted or ratified.'[17] Thus it was not the states that created the union; the people as a whole came first. And this gives a completely different sense to the Constitution. Wills says that Lincoln 'altered [the Constitution] from within, by appeal from its letter to the spirit.'[18]

To be sure, Lincoln did not invent the idea of the primacy of the union over states' rights, much less present it for the first time in the Gettysburg Address. Along with slavery itself, it had been the central issue in the debates leading up to the Civil War, and the founding role of the Declaration had been championed eloquently by Daniel Webster in the 1830s.[19] But at Gettysburg Lincoln compressed these ideas into a few words, and the success of his speech etched them for good into the national consciousness of Americans.

Thus Lincoln's speech is important here not because of the originality of its ideas, but as a prime example of social or communal narrative with a practical purpose. As Wills shows, this purpose lies far beyond rallying the people to continue to fight and win the war. It *is* about the future, but its intent is far more serious, and so it must engage the past as well. As Wills puts it: '… the crowd [at Gettysburg that day] departed with a new thing in its ideological luggage, that new constitution Lincoln had substituted for the one they brought there with them. They walked off … into a different America. Lincoln had revolutionized the Revolution, giving people a new past to live with that would change their future indefinitely.'[20]

I have examined Lincoln's speech as an illustration *par excellence* of the way the past functions as 'society's memory' in the context of a social narrative. Here the past is anything but static; it is changeable, a vital and important dimension of the life of any community. Above all, it is a serious matter, something to be sruggled over. As we have seen, the basic motive of the social narrative at this level is practical, and its primary orientation, despite its preoccupation with the past, is really the future.

It is against this background that we must now assess the character of historical narrative proper, the sort we associate with our historical discipline today. By contrast to the future-orientation of social narrative, historical narrative seems explicitly occupied with the past for its own sake. And its underlying motive could be characterized as strictly cognitive, not practical. Moreover, the *cognitive* interest is largely equivalent to the *critical* interest. The historian often examines with a critical eye precisely those versions of the social past that exist in the collective memory. These are now regarded as myths and legends, in the pejorative sense of those terms, and the practical – or *partisan* – character of their origin is what comes under critical scrutiny. By contrast, the historical narrative is meant to be unbiased, objective, purely 'scientific.' Whether it can ever achieve this goal is not at issue here; we are speaking of its underlying values and intentions. It is important to note that the value of objectivity extends beyond securing 'facts' by attending to sources. As Ruesen has shown, the critical use of sources, already enshrined as a value in Enlightenment historiography, was not enough for the historicists of the nineteenth century: the interpretation of facts and their incorporation into narratives were likewise supposed to live up to rigorous methodical standards.[21]

Thus historical narrative differs from social narrative not only by its past orientation and its aim of objectivity; it lies beyond the social narrative not so much by being a different story as by reflecting back on social narratives with an eye to critical evaluation and, if necessary, revision. This formulation gives a first approximation to an account of how historical narrative relates to the first-order practical narratives of historical 'reality,' whether they be individual or communal. We begin to see specifically what it means to assert a 'continuity' between historical narrative and what it is about.

These considerations permit us to answer at least one critical assessment of the continuity theory.[22] If historical narrative is not imposed on historical reality as an alien and distorting medium, but rather shares with it the narrative form, then perhaps this means that the only task of the historian is to *report* or even *repeat* the content of those implicit and explicit narrative self-descriptions of historical agents. This would place the continuity theory very close to Collingwood's idealist theory of re-enactment and the classic concept of *Verstehen*.[23] The task of the historian would be to grasp the thoughts or reasons behind the actions of historical agents, to understand events as they understood them. In narrative terms, this would mean to retell the story such agents told themselves and others, implicitly or explicitly, in order to act as they did.

But there are well-known difficulties associated with Collingwood's idealism, and they would be found in this narrative version of it as well. It would give us a simplistic caricature of what historians actually do. It would ignore the critical and revisionist motive behind historical narrative. The historian needs to go beyond any single description of an action or event by comparing it with other descriptions of the same event, thus having to deal with problems of identification across multiple descriptions (compare Luther's description – or 'story' – of what he did in 1517 with the Pope's description of the same events). More important, this notion ignores what many theorists, from Dilthey to Danto, have seen as history's unique cognitive advantage, that of hindsight. The retrospective glance of the historian permits her to view the self-descriptions and practical narratives behind historical actions in light of their actual outcomes. The ironic disparity between the envisioned or intended and the actual consequences of an action is very important for the historian.

It is true that the 're-enactment' or revival of agents' own narratives is a vital part of this retrospective comparison, a point often forgotten by those like Danto who stress the retrospective point of view.[24] The historian is not only capable of, but usually *cannot avoid* viewing past events in light of their actual consequences. The interest and value in historical accounts often lies precisely in retrieving a perspective on events which has been lost to us because of our hindsightful wisdom. Luther did not plan, envision or desire the Protestant Reformation that was unleashed by his actions. Seeing such a person's action just as he saw it, deliberately blocking out all that we know of what came later, must surely be one of the most difficult feats of the historical imagination. It should be pointed out in passing that this problem is not confined to narrative history, nor is it only a problem of 'empathizing' with a remote historical individual. A history of mentalities or concepts, of the sort inspired by Foucault, similarly seeks to block out the vision of hindsight by refusing to see past events (including theories or writings) merely as precursers of what came later. Thus, for example, Newton's theories must be viewed in light of the religious cosmology of his time, not merely as precursors of the physical theories of our own day.

There is another sense in which this operation of re-enactment is important. Many historians, especially today, see their activity as one of redressing certain imbalances, of rediscovering or retrieving what has been lost, forgotten or covered over. The lives of those excluded from the stories of the past or relegated to their margins, those selected out of the standard narratives of both historical agents and later historians, such as women and minorities, are to be reinstated in our historical consciousness. And the first task is to give them back their own voice if possible, let them tell their own stories just as they are articulated in diaries, speeches, sermons, court testimony, folk art or other expressions. Here too the historian must in a sense block out hindsight and efface himself before the the voice of the past.

Important and necessary as these operations of retrieval are, one should not conclude from them that they make up the whole work of historical narrative. Any account that would restrict its role to that of repeating the narratives of historical agents would overlook the most important thing: history has its *own* story to tell; its purpose is to generate a narrative which is different from and goes beyond any of the narratives it may incorporate. Even in re-telling stories, the historian makes a contribution, since she must choose which stories to re-tell. And these stories are typically integrated into the larger context provided by the broader view, and of course the retrospective view, of the historian. This in turn introduces the critical perspective from which the individual and social narratives are evaluated in objective and unbiased terms. To put it crudely, the historian is interested in events *as* they were lived and described by participants, but with a view to contrasting those descriptions with events as the 'really were,' that is, as they are now viewed by the historian and integrated into his or her own narrative.

Thus the relation of narrative history to the narrative 'material' with which it works is a complex one. Its link to the continuum of individual and social narratives is threefold: First, it reflects back on them in order to retrieve them as they are, revive or re-enact them as closely as possible to how they were experienced by participants. Second, it reflects critically on them by comparing them with each other and with their real, as opposed to their intended, consequences. Third, these narratives are incorporated into a *new* story which is that of the historian himself. Once this last point is recognized, the claim of the continuity theory is borne out, namely that in an important sense history does the *same thing* as individual and social narrative, even though it may do it at a different level and by different means.

But it also does it for different *reasons*. It was suggested earlier that historical differs from social narrative in motivation and in temporal orientation. Social narrative is practical, we said, and is oriented toward the future, while historical narrative is purely cognitive and is occupied only with the past. Do these differences constitute an irreconcilable disparity which would undermine the point of the continuity theory? Can we really compare the activity of telling a 'likely story' that will contribute to getting things done, with that of recounting the past 'wie es eigentlich gewesen'? Is there not all the difference in the world, indeed, between telling a *useful* story and seeking the *truth*?

In fact I think this is probably a false dichotomy and at most a matter of emphasis. Story-telling at all levels of practical life, provided it is not admittedly fictional, is always constrained by truth. Falsehood may in some circumstances, and in the short run, be more useful than the truth, but ultimately the truth is more practical. Our story-telling must come to terms with the world as it is, not as we wish it were. 'Making sense' cannot be separated from 'being true.' 'Getting the story straight' is a value at all levels of individual and social existence, not just at the level of science or scientific history. Methods of evaluating sources, notions of objectivity, injunctions of impartiality, are ways of trying to do better what we often do poorly in ordinary life. Thus the historian's concern with the truth about the past is ultimately practical, too. And to say that it is practical is to say that it has value for the future.

With this account we can thus conclude our response to the second of the two challenges to the continuity theory. This theory can do justice to the specificity of historical narrative while still relating it closely in structure and motivation to the practical narratives of individual and social life.

Notes

1 Ruesen (1983), p. 52.
2 Ricoeur (1983), p. 85.
3 See Carr (1986), Ch. 2 and *passim*.
4 Mink (1987), p. 60.
5 White (1987), p. 44.
6 See Ricoeur (1992), especially the fifth and sixth studies, pp. 113–68.
7 Dilthey (1968), pp. 277–8.
8 Hegel (1976), p. 83.
9 Kulke (1994/95).
10 Lévi-Strauss (1966), pp. 234–5, and Eliade (1956), Chs 3 and 4.
11 Miller (1995).
12 Kosellek (1979), 'Historia Magistra Vitae' and other essays.
13 See Lévi-Strauss's attack on Sartre (1966), pp. 248–9.
14 Wills (1992).
15 Ibid., p. 55.
16 Ibid., p. 53.
17 Ibid., p. 129.
18 Ibid., p. 38.
19 Ibid., p. 130.
20 Ibid., p. 38.
21 Ruesen (1993), pp. 61f.
22 Norman (1991), p. 124
23 Collingwood (1969) pp. 282ff.
24 See Danto (1965).

References

Carr, David (1986), *Time, Narrative and History*, Bloomington: Indiana University Press.

Collingwood, R.G. (1969), *The Idea of History*, London: Oxford University Press.

Danto, Arthur (1965), *Analytical Philosophy of History*, Cambridge: Cambridge University Press

Dilthey, Wilhelm (1968), *Gesammelte Schriften*, Vol. VII, 5th edn, ed. B. Groethhuysen, Stuttgart: B. Teubner.

Eliade, Mircea (1956), *Cosmos and History*, New York: Harper and Row.

Hegel, G.W.F. (1976), *Vorlesungen ueber die Philosophie der Geschichte*, Frankfurt: Suhrkamp.

Kulke, Hermann (1994/95), 'Geschichtsschreibung als Heilung eines Traditionsbruches?', Report No. 1/94 of the 1994/95 research group on Historische Sinnbildung, Center for Interdisciplinary Research, University of Bielefeld.

Kosellek, Reinhard (1979), *Vergangene Zukunft: Zur Semantik geschichtlicher Zeiten*, Frankfurt: Suhrkamp Verlag.

Lévi-Strauss, Claude (1966), *The Savage Mind*, Chicago: University of Chicago Press.

Miller, Klaus E. (1994/95), 'Praehistorisches Geschichtsbewusstsein', Report No. 1/94 of the 1994/95 research group on Historische Sinnbildung, Center for Interdisciplinary Research, University of Bielefeld.

Mink, Louis (1987), 'History and Fiction as Modes of Comprehension' in Brian Fay, Eugene O. Golob and Richard T. Vann (eds), *Historical Understanding*, Ithaca, NY and London: Cornell University Press.

Norman, Andrew P. (1991), 'Telling it Like it Was: Historical Narratives on their Own Terms', *History and Theory*, 30 (2).

Ricoeur, Paul (1983), *Temps et Récit*, Vol. I, Paris: Seuil.

Ricoeur, Paul (1992), *Oneself as Another*, trans. K. Blamey, Chicago: University of Chicago Press.

Ruesen, Joern (1983), *Historische Vernunft*, Göttingen: Vandenhoeck and Ruprecht.

Ruesen, Joern (1993), *Konfigurationen des Historismus*, Frankfurt: Suhrkamp.

White, Hayden (1987), 'The Question of Narrative in Contemporary Historical Theory', in Hayden White (ed.), *The Content of the Form*, Baltimore, MD: Johns Hopkins University Press.

Wills, Gary (1992), *Lincoln at Gettysburg: The Words that Remade America*, New York: Simon and Schuster.

Chapter 6

The Social Location of Scientific Practices[1]

Hugh Lacey

The mainstream tradition of modern science denies the *historicity* of scientific practices. It denies that their *character* changes, and must change, in fundamental ways that arise historically, and that are responsive to and shaped significantly by varying circumstances. In Section 1 I will identify several presuppositions that have been commonly used to support this denial: most notably, that the object of science is ahistorical and its methodology essentially unchanging, and that the character of basic scientific methodology is not dialectically linked with applied science. In Section 2 Kuhn's rejection of the presuppositions about the object and unchanging methodology of science will be endorsed, thus providing support for what Margolis has referred to as a 'remarkable (but somewhat muffled)' version of the historicity of science.[2] Then, in Section 3, drawing upon a detailed analysis of a contemporary controversy between agrobiotechnology and agroecology, I will move beyond Kuhn and also reject the other presupposition that there are no dialectical links between methodology and application.

Specifically I will argue that the character of scientific practices reflects mutually reinforcing relations with the social location in which they are conducted, that is, relations with the value-outlooks of their practitioners and their enabling institutions, and the interests that will be served through applications of their products. It is a small step from this to endorsing an 'unmuffled' version of the historicity of scientific practices, one that admits that variations in the character of scientific practices may be dialectically linked with historical and cultural variations in the realm of daily life and experience and in the structures of social practice.

1 Scientific Practices as Lacking Historicity

Let me briefly recapitulate a story that articulates an important part of the self-understanding of the modern scientific tradition, and that has often been re-told. Science has a *history*, a history of progress: of growth, accumulation and refinement of scientific knowledge, and of elimination of error. It is a history in which *methodology* plays a central role. Provided only that scientific practices are kept free from outside interference and nourished from time to time by the input of creative genius, methodology ensures the continued unfolding of the progressive development of science. Scientific methodology is systematic and empirical, rooted in experiment and measurement. It prescribes that empirical data be brought to bear upon theories that, using the resources of mathematically articulated lexicons, posit representations of phenomena and their underlying order and law. Apart from refinements of detail, scope and precision, scientific methodology does not change. Thus, the 'scientific revolution' of the sixteenth and seventeenth centuries marks the effective *beginnings* of science (anticipated only by scattered fragments of scientific knowledge), not a fundamental *change* in the methodological character of scientific practices.

Since then, the story continues, the cognitive (epistemic) credentials of scientific methodology have been certified and repeatedly vindicated. Technological success that has been informed by scientific knowledge has been one source of the vindication. Another has been the knowledge and understanding of 'the world' ('the natural world,' 'the material world') – of natural laws, and of things, events, states of affairs, phenomena, structures and their underlying components, processes and interactions – that have accumulated and been refined, and whose compass continues to expand, bounded only by the limit of a 'complete account' of 'the material world,' one that in due course would encompass all phenomena.

The story admits of competing versions with different emphases about, for example, the primacy of theory or experiment and the significance of applied science. In all versions, however, the tale of progress attends principally to such matters as theories that have been developed, available data, technical possibilities for experiment and measurement, methodological matters, and the (creative) inputs of individuals (or groups) of scientists. That way the 'rationality,' 'universality' and 'objectivity' of the cumulative and developing process are able to be emphasized. There is a place in the story for social, economic and political factors: sometimes the interests of utility lead to a focus on a particular object of inquiry and, more generally, the rhythms and organization of scientific research depend upon the availability of the appropriate material resources and social conditions. Even so, the fundamental dynamic of scientific progress is 'rational': science is progressively, and with ever greater refinement, gaining knowledge of objects of 'the material world.' Social (and hence historical) factors may lead to giving priority, even urgency, to gaining knowledge of specific instances of these objects; but gaining knowledge of them contributes to the overall accumulation and refinement of knowledge of 'the material world.' It is the accumulation and refinements that matter most. The rest, including the temporal order in which objects are investigated, is incidental. Nowhere in the story does the character of scientific practices change in fundamental ways; they do not exhibit historicity.

1.1 Suppositions Supporting the Denial of Historicity

At the root of this denial of historicity are the following three suppositions:

1. Science aims to gain a kind of understanding that is expressed in theories that match ever more completely and accurately an ahistorical object, 'the material world,' whose underlying order (laws; and structures and their components, processes, interactions) is *ontologically* independent of human actions, desires, conceptions, observations and investigations.[3]
2. The methodology of modern scientific practices (subject only to refinements of precision, scope and the like, but not to any fundamental change) enables us progressively to gain understanding of this ahistorical object – so that there is no deep historical dialectic of methodology and object of inquiry, and so that the questions posed in basic science (while they might depend on the results of previous inquiries and the availability of instrumentation and appropriate mathematical and conceptual resources) do not concern objects insofar as they are historically variable, socially located, or playing integral roles in human practices.
3. The acceptability of scientific theories depends only on considerations involving their features and their relations with empirical data of selected kinds.

Clearly, and consistent with supposition 1, the actual arrangements of material objects in our vicinity are not *causally* independent of human affairs. These arrangements may be consequences of scientific applications; so much so that, although science supposedly lacks historicity, in virtue of its applications it has become nevertheless a historical agent of extraordinary importance. Indeed, it has been held, the very success of science in informing technological developments is explained in terms of its having gained sound understanding of 'the material world.'[4] According to the story under review, the historical agency of science may account for the ready availability in the advanced industrial countries of the social conditions and material resources required for the pursuit of science. Moreover, nowadays a good deal of research depends upon the availability of instruments that are the products of the most advanced and sophisticated technology, whose availability is itself made possible by scientific developments. This means that the historical agency of science functions as an 'instrumental partner' of scientific research, one that enables the methodology to be deployed in a more refined way (for example, enabling us to obtain greater precision in measurement and to explore hitherto inaccessible spaces). It feeds back so as to serve the cognitive (epistemic) interests of gaining scientific understanding (a 'happy coincidence' of social practical interests and knowledge-gaining interests[5]), but it leaves the fundamental character of scientific methodology essentially unscathed. Thus, the denial of historicity also involves the supposition:

4 The historical agency of science (exercised through its applications) is only a consequence and an instrumental partner of successful scientific practice; it is not a dialectical partner, one that feeds back so as to influence the fundamental methodological character of these practices.

Affirming the historicity of science involves denying suppositions 1 and 2: denying that the object of scientific investigation is ahistorical, and maintaining that there is a dialectic between methodology and object of inquiry. It is deepened by also denying supposition 4, affirming that there is a dialectic between methodology and the practices of socially applied science. Kuhn has made a compelling case for the denial of suppositions 1 and 2, though he seems to accept supposition 4.[6] Before addressing Kuhn's argument, let us extend the story being re-told a little further.

1.2 Do Soundly Accepted Theories Represent the 'Material World'?

According to our story, *objects as grasped in the practices of basic science*, that is, *objects as represented in soundly accepted theories*, are (approximately) identical to *objects as they are in the underlying order of the material world.* Scientific practices, and the modes of interaction and thought that constitute them, enable us to grasp things as they *are* in the ahistorical 'material world,' abstracted from the context and conditions of our investigations, and indeed from all human related contexts. Since scientific practices are themselves historical, how can this be so? Methodology is the key to the answer; but how is it that a methodology deployed within a historical practice can enable us to grasp the ahistorical?

It is able to do so, a common response goes, in virtue of the character of and relationships between *theory* and *empirical data*. To elaborate: theories developed in scientific practices deploy carefully expressed posits (and models), typically in mathematical form, about underlying (non-apparent) structure and its components, process, interaction and law; so that theories abstract the phenomena investigated from their places in the social order, in daily life and experience, and even in scientific practices themselves. And data are sought out and reported, and the conditions in which they may be obtained are often created, in the course of experimental and measurement practices. Relevant data, obtained from observing phenomena of which a theory is proposed to provide understanding, meet the condition of intersubjectivity (and, where possible, replicability), and quantitative and experimental data are of special significance. Then understanding of objects of 'the material world' is expressed in soundly accepted theories.

A theory is *accepted* if its posits (pertaining to certain domains of phenomena) are put into the stock of established scientific knowledge, the stock of those posits judged to be such that further investigation or testing of them would produce at most refinements of accuracy and scope.[7] A theory is *soundly accepted* (of the phenomena of a specified domain) if it satisfies certain criteria, that is, if it manifests the *cognitive values* highly in relation to the available data from this domain – if it has specified characteristics (for example, consistency, simplicity), relations with other accepted theories (for example, inter-theoretic consistency, consilience), relations with displaced theories (for example, being a source of interpretive power of the strengths and weaknesses of a displaced theory), and most importantly relations with available empirical data (for example, empirical adequacy, explanatory and predictive power).[8] Theories that have been soundly accepted of specified domains have also reliably informed numerous practical (technological) applications.

What legitimates the move made in the story from (a) 'T manifests the cognitive values highly with respect to D' to (b) 'T represents (matches) order of the "material world" underlying D'? (T is a theory and D is a domain of phenomena.) One might respond: Is it not obvious, given that T represents the phenomena of D in abstraction from the relations they may have with human and social affairs, in terms of their being generated from the underlying order, and that the sound acceptance of T depends *only* on judgments of the manifestation of the cognitive values in T with respect to D? Moreover, that the move has been frequently and casually made throughout the course of modern science suggests that it is taken to be obvious. In what else could the cognitive value of T consist other than a match with parts of the 'material world'? Well, it could consist in confirming (c) 'T encapsulates well the possibilities of phenomena of D in so far as they derive from the generative power of the underlying order (their '*material possibilities*'),' or (equivalently) in grasping these phenomena *qua* abstracted from their human and social contexts.

Elsewhere I have argued that the move from (a) to (b) is not mediated by (d) 'T reliably informs technological applications';[9] but the move from (a) to (c) is supported by (d). Often the move from (a) to (b) is made against background commitment to materialist metaphysics, which may be considered as a suitable elaboration of the posit that all phenomena are lawful or that all possibilities are material possibilities: the 'material world' – the ahistorical order underlying things – *really is* such that it can be matched by (and only by) the kinds of posits put forward in modern scientific theories. Were there a sound *a priori* case for materialist metaphysics, this might be compelling. But today, for the most part, those who espouse materialist metaphysics do so on the ground that it is an extrapolation from established scientific understanding and the direction of its expected growth. Then, if materialist metaphysics provides the ground for the move, the question is begged.

2 Kuhn's Account of the Historicity of Science

Kuhn maintains that there is nothing in the character of scientific practices that justifies the move from (a) to (b),[10] and that, furthermore, attention to the actual history of science suggests that the move would be clearly unjustified. In the history of science, he maintains, we do not find steady accumulation and refinement, but instead periods of fundamental discontinuity in the character of scientific activity – discontinuities (for example) in what is considered a theory worthy of provisional investigation, in what are the appropriate phenomena to investigate for the sake of gaining empirical data (and in the descriptive categories of the data) that are to be fitted by theories, and in what kinds of posits are taken to be central for shaping scientific investigation.

2.1 Soundly Accepted Theories are Developed and Consolidated under a 'Strategy' ('Paradigm')

According to Kuhn, if theory and empirical data are taken to be *the* major elements of scientific methodology, no sense can be made of the actual history of science. Kuhn proposed a third element: *paradigm*, of which

I will consider just one aspect: within a paradigm, research is conducted under a *strategy*.[11] A strategy specifies *constraints* upon theories that are taken to be admissible for provisional consideration (and possible eventual acceptance), and (reciprocally) criteria upon the kinds of empirical data (and the phenomena from the observation and measurement of which they are obtained) that are *selected* as those appropriate for being brought into the appropriate relationships with theories. These are the kinds of data needed for testing and selecting among provisionally entertained theories, and those that describe phenomena so as to enable their explanations and the encapsulation of their possibilities. Admissible theories may be constrained, for example, to be formulated with the resources of a specified lexicon (for example, the teleological/sensory categories of Aristotelian physics, or the mathematical/mechanical ones of Galilean physics), and the data may be selected (generally subject to the condition of intersubjectivity and, where appropriate, replicability) in virtue of (for example) being representative of phenomena of daily life and experience, or of pertaining to experimental and measurement practices.

Given this third methodological element, we are able to identify two key (logically distinct) moments of choice: choice of strategy to adopt in research practices, choice of theory to accept or reject. Choice of theory is, then, in the first instance choice among provisionally entertained theories that fit the constraints of the adopted strategy.[12] When properly made, it involves judgment about which one of them best manifests the cognitive values with respect to the available data, about whether the available data are sufficient and about whether the manifestation meets high enough standards for accepting the theory of the relevant domains of phenomena.[13] Accepted theories encapsulate soundly certain kinds of possibilities that these phenomena permit. (Successful application testifies to this.) So adopting a strategy involves identifying the kinds of possibilities desired to be encapsulated; accepting a theory involves identifying (typically through consolidating posits about how to actualize them) the genuine possibilities of these kinds.

In the light of the Kuhnian insight, our initial story can be reinterpreted or (more accurately) replaced by a narrative of research conducted under a particular set of strategies – that I call *materialist strategies* (MS) – that have been adopted virtually exclusively within the modern scientific tradition.[14] MS incorporate the core methodological elements cited in the story. Under them, theories are constrained to those with the lexical and mathematical resources to be able to formulate posits of underlying order – structure and its components, process, interaction and law, where laws express relations among quantities. These theories identify the possibilities of phenomena in terms of the generative power of the underlying order (the ones I have called 'material possibilities'), in abstraction from any place they may have in human experience and practical activity, from any links with social value and with the human, social and ecological possibilities that they might also admit. The 'material possibilities' of phenomena include possibilities that are identical with possibilities for technological application. Reciprocally, under MS, data are selected (subject to intersubjectivity and replicability) so that their descriptive categories are generally quantitative, devoid of the categories of intentionality and value, applicable in virtue of measurement, instrumental and experimental operations.

Research conducted under MS has been extraordinarily successful: it has generated and continues to generate soundly accepted theories of a great variety of phenomena; and these theories have been the source of numerous and varied technological applications. Kuhn has little interest in applications. And, as I have said, he does not take the success of MS in producing soundly accepted theories to show that these theories match the ahistorical 'material world.' Instead, for Kuhn, it establishes that the world can be (to a marked extent) well matched to, or become amenable to grasp within, the categories of the lexicons deployed under MS.[15] I add that 'material possibilities' of things are successfully identified under these MS; and, in opposition to those who adhere to materialist metaphysics, I caution that there is no reason to believe that the possibilities of things are exhausted by their material possibilities.

Within the Kuhnian picture, the object of scientific inquiry is phenomena *qua* grasped under a strategy. Since a strategy is a methodological innovation of scientific practices, this object is not ahistorical. For Kuhn himself, the aim of science is to solve puzzles whose very definition is strategy-bounded. In the final analysis the very questions posed in scientific inquiry are not about the 'material world,' but about the power of a strategy to grasp phenomena. It follows that suppositions 1 and 2 (see previous section) are not sustainable. On my additional gloss, the aim of science is to gain understanding of phenomena, and this includes encapsulating the possibilities that they allow.[16] But phenomena allow many and varied kinds of possibilities, not all of which can simultaneously be co-actualized or even co-investigated – so actual scientific investigation opts to pursue certain classes of *valued* possibilities, generally those valued for the sake of application, whose realizability and possibility of being investigated is historically conditioned. Thus the unsustainability of suppositions 1 and 2 is reinforced.

2.2 'Fruitfulness' as Ground for Adopting a Strategy

What about supposition 4? It could not be sustained, if (for example) the (rational) grounds for adopting a strategy include that it gives rise to applications of special interest for those holding a particular value-outlook. So we must attend to the grounds for adopting a strategy.

According to Kuhn, a strategy is adopted for the sake of defining and solving puzzles, or (as I prefer to put it) for the sake of generating theories and acquiring appropriate empirical data so that theories can come to be accepted in virtue of manifesting the cognitive values highly. Then, a strategy is worthy of adoption only if it is demonstrated to be *fruitful* – actually to be, and continuing to be, a source of theories that come to be soundly accepted of certain domains of phenomena. A fruitful strategy, adopted in the first instance following an exemplary achievement, enables investigation to take place in the relevant field;[17] and, for Kuhn, so long as a strategy remains fruitful, research should be conducted *exclusively* under it. Within the scientific tradition, he maintains, fruitfulness is sufficient, as well as necessary, for the adoption of a strategy. *Normally* a currently fruitful strategy is in place. Then, engaging in scientific research implies adopting it – so that normally questions about adoption of strategy are neither controversial nor addressed explicitly within the scientific community[18] – until such time as the limits of its fruitful unfolding are reached. Such limits become apparent when *anomalous* phenomena (which have become considered important for the unfolding research) are identified: phenomena that cannot, after prolonged and skilful investigation, be fitted into theories that both meet the constraints of the strategies and manifest the cognitive values highly, but at best into theories that retain empirical adequacy at the price of increasingly diminished manifestations of such other cognitive values as predictive and explanatory power and keeping *ad hoc* hypotheses to a minimum.

On Kuhnian views, strategies and the lexicons they bear are human creations; and a soundly accepted theory is one that succeeds in fitting certain phenomena of the world into the structured lexicon of a strategy. So it is expected that any strategy will have limits, that its fruitfulness will eventually become exhausted. (Any one kind of strategy will fail to encapsulate various kinds of possibilities of phenomena.) When the limits of an established strategy are reached, and – according to Kuhn – (allowing a certain latitude of judgment about when they are reached) only then, does the scientific tradition license the search for another strategy; and then the search is for a new strategy that can grasp the anomalies of the old one. At such (*revolutionary*) moments most of the old constraints are lifted, conflicting perspectives are engaged and there is much trial and error, until such time as a new strategy emerges (in a new exemplary achievement that offers promise of further fruitful developments) that enables the grasp of the old anomalies.[19] Then that strategy comes to demand the allegiance of the scientific community.

Kuhn intends his picture to be both descriptive (under idealization) of the history of science, and normative for scientific practice. Indeed, normally it does suffice for scientific research to proceed under a single strategy, *provided that* one accepts that what count as scientifically interesting phenomena are defined within the unfolding tradition of science, and that the aim of science is to resolve puzzles about them or to come to accept (soundly) theories of them. Proceeding in this way enables there to be successful research, practically ensures that empirical considerations will eventually lead to clear demarcation of the limits of the strategy, and keeps a measure of continuity – through the special role accorded to anomalies of old strategies – across the 'revolutionary' divides that separate the periods of hegemony of succeeding strategies. Note that an argument cannot be extracted out of this that the new strategy is the only one that could have developed as successor to the old one.[20] Within the Kuhnian picture, there are elements of radical contingency: that any successor at all will actually emerge, and if one does, what its specific character will be. The emergence of a new strategy may be influenced causally by all sorts of 'extra-scientific' factors (religious, metaphysical, cultural), but what matters, what legitimates the adoption of the strategy, is that it generates theories in which the anomalous phenomena can be grasped and which define new puzzles. If the aim is to solve puzzles about scientifically interesting phenomena and to introduce new ones to be solved, that is enough.

Kuhn has provided a brilliant account of the transition from the hegemony of Aristotelian to that of materialist (Galilean) strategies[21] as well as some less developed accounts of other 'revolutionary' transitions.[22] Following the former transition, few products of Aristotelian science have remained in the generally accepted stock of knowledge. With the hindsight of developments under MS (including new data, greater sensitivity to the role of certain cognitive values, and higher standards for estimating the degree of manifestation of the cognitive values in theories), it became apparent that Aristotelian physical theories were soundly accepted of very few phenomena.[23]

Some of Kuhn's critics think that his view entails that, with the eventual anticipated surpassing of MS as framers of research, few of its products will remain in the stock of knowledge. (Thus they accuse Kuhn of a kind of relativism that seems manifestly unacceptable when we think of the discoveries of modern science and their applied successes.) But Kuhn's view does not entail this. Under MS, numerous theories have been soundly accepted of countless domains of phenomena. These theories encapsulate well an increasing number and variety of the material possibilities of phenomena; and – while acknowledging the truism that empirical methodologies cannot provide certainty – there is no reason to hold that subsequent developments of the tradition will lead to removing from the stock of knowledge the confirmation of these possibilities. Similarly, there is no reason to anticipate (for example) that the atoms of modern atomic theory will go the way of the four terrestrial elements of Aristotelian physics, at least if we consider atoms to be the constituents of molecules with capacities for generating specified effects in specified (experimental and technological) spaces. Subsequent research may lead to their refinement and elaboration, but given how soundly accepted atomic theory is, not to their rejection. I have followed Cartwright's language here,[24] and her claim that established scientific knowledge is largely of capacities of objects: that they tend to have certain effects under specified (typically experimental) conditions, without the further supposition that such capacities (rather than others they might also have) will be exercised significantly in all ('natural') situations. Gaining such knowledge of capacities of objects does not ground the supposition that knowledge of the 'material world' – of the world as it *is* independent of its relations with human beings – has been gained. Only idle skepticism would cast doubt on the existence of atoms today: there are atoms in the world 'that we live in' and 'that we have investigated,' and we know their capacities as exercised in various experimental and technological spaces and also (no doubt) in many spaces, not of human causal origin, in which there is no (relevant) human causal involvement. Kuhn's picture fits easily with many kinds of scientific realism. But the 'world that we live in' is not the 'world as it *is* independent of its relations with human beings.'

Thus, it is consistent with the Kuhnian picture that, under MS, we gain accumulating knowledge of *'material possibilities' of phenomena*. But, one might ask, is this really any different from accumulating knowledge of *the ahistorical material world*? It is, and the difference is of central importance. In the first place, the latter idiom, unlike the former, is usually linked with the view that all possibilities of phenomena are (in the final analysis) material possibilities, and in particular with materialist reductive accounts of human cognitive (rational) and moral capacities. Secondly, the material possibilities of phenomena are those possibilities that are encapsulated by the generative power of the underlying order posited of the phenomena; they are constituted as such within scientific practices conducted under MS. Some of them are realized in, and realizable only in, experimental and technological spaces of human creation (having been posited as the possibilities of these historically bounded spaces). Others are realized in spaces whose underlying *causal* order, as represented under MS, has no relevant human involvement, but where that causal order is posited, generally drawing upon the resources of theories accepted of experimental spaces, following scientific observation aided by instruments themselves authenticated in the course of experimental and technological practices. (In some spaces there can be good reason to hold that their possibilities are exhausted by their material possibilities.) This causal order is constituted in the course of scientific practices as a projection from experimental and technological practices; there is no basis here to infer to the features of an underlying order that is *ontologically* independent of human beings.[25]

For a theory developed of a domain of phenomena under MS, 'T manifests the cognitive values highly of D' implies 'T soundly encapsulates material possibilities of D.' These propositions become established at a particular time. Nevertheless, once established, especially if further vindicated by the success of practical applications, there is no *general* reason to expect that they will become vulnerable to refutation in the light of outcomes of research under different strategies, either current alternative strategies or future ones. Strategies change, and so the fundamental character of ongoing scientific investigation changes, but that permits a permanent residue of knowledge to remain, a residue that may or may not become rearticulated (as a particular case or as an approximation) under a subsequent strategy. Historicity of scientific practices does not imply the historical relativity of scientific knowledge. It does make likely, however, that quite a bit of what is taken to be established scientific knowledge (but with insufficient scrutiny of the degree of manifestation of the cognitive values) will come to be recognized as not properly part of the permanent stock of knowledge. And it fits easily with the historical (and cultural) relativity of interest in applying particular items of scientific knowledge.

As more material possibilities become soundly encapsulated in theories, the greater is the range of technological possibilities opened up, a matter with profound social implications. For Kuhn, technological application remains principally a *consequence* of scientific developments, and also a source of additional empirical data to bring to bear on theories, especially by way of the instrumental partnership referred to in Section 1. That there is widespread technological application, and that it is desired, are not among the (rational) grounds for adopting MS; those grounds are (normally) solely connected with fruitfulness, and also (at 'revolutionary' moments) with being able to grasp the anomalies of the old strategy. Through this complex and subtle narrative Kuhn endorses the historicity of science: denying suppositions 1 and 2, while retaining supposition 4, and thus preserving an essentially internalist narrative of the history of science.

3 The Role of Applications

'Application' refers to two interacting and not sharply separable roles that scientific theories can play in social practical life. A theory may apply *to* significant phenomena of daily life and experience; and it may be applied *in* practical activity. It applies *to* those phenomena when it is used, by way of representing them with its categories and principles, to provide understanding of them. 'Applying *to*' involves identifying (modeling) phenomena as characterized using everyday categories with phenomena as represented in the theory. A theory is applied *in* practical (often technological) activity when its posits inform such concerns of practice as the workings of things, means to ends, the attainability of ends, and the consequences of realizing the possible.[26]

In Kuhn's picture, applications are important to the unfolding of the scientific tradition only as enticement for the provision of the social, material and instrumental requirements of the conduct of research. Its credibility depends on dissociating the value, conduct and *character* of scientific practices from social and moral evaluations of applications of the knowledge they produce.

I will now offer an alternative picture in which applications (*to* and *in*) are more central than Kuhn admits. In it, particular strategies are adopted rationally (in part) – subject to fruitfulness remaining a necessary condition of their adoption – *because* they can be expected to give rise to *certain kinds* of applications.[27] Phenomena are in fact (and should be) brought to the attention of basic scientific investigation, *not only* from the scientific tradition's own unfolding (as Kuhn holds), *but also* from the realm of daily life and experience and social practice, from the 'world in which we live.' Science aims to provide understanding of phenomena and, in doing so, where appropriate to make sense of our experience and to inform our social practices. Strategies worthy of adoption should normally produce theories applicable *to* phenomena significant for current daily life and applicable *in* current social practices – though normally and desirably (for substantive and methodological reasons) the reach of scientific investigation should not be limited to phenomena involved in these applications. Many significant phenomena of daily life and social practice are not fixed across historical change and cultural variety so that depending on the desired applications, different strategies may be needed. If so, supposition 4 could not be sustained.

In order to provide detail and credibility for this alternative picture, I will show that, for some significant phenomena, competing (fruitful) strategies are possible. I will focus upon phenomena encountered in farming practices. For them, which of the strategies becomes adopted in research depends upon the social location of the investigator, and upon the way in which applications are valued from this location. Different social locations (on the one hand, the neo-liberal global economic project; on the other, grassroots movements of poor farmers) lead to the adoption of largely different (competing) strategies.

3.1 Do Materialist Strategies Suffice to Shape Research?

The modern realm of daily life and experience is unintelligible apart from the applications of knowledge gained under MS, since it has been shaped to a great extent by identifying and realizing novel material possibilities of things. That provides a good reason for MS to be adopted in the scientific community.[28] But, the possibilities of natural phenomena encountered in daily life and social practice are not reducible to their material possibilities, those they have in virtue of the generative power of their underlying structure (and its components), process, interaction and law. Why, then, prioritize material possibilities in the investigation of natural phenomena? Why not attempt to shape and adopt strategies under which other classes of their possibilities might be identified, for example those they have in virtue of their places in human life and experience and social/ecological systems? Why, for example, prioritize investigating seeds so as to identify the possibilities open to them under the genetic modification procedures of current biotechnology, rather than those they have in virtue of their place in productive and sustainable agroecological systems?

The following answer might be part of the continuation of the story told in Section 1: non-material possibilities of objects (phenomena) supervene on their material possibilities. The realization of a material possibility (where human interactions with natural objects are involved) may be identical to the realization of a social/ecological possibility. Successfully producing genetically modified seeds and reaping a harvest from them, for example, is also at the same time (under current socioeconomic conditions) furthering the process in which seeds become commodities.[29] But the class of material possibilities can (in principle) be identified simply in terms of the generability of each of its members from the underlying order. Some material possibilities may also be identified *qua* social possibilities – but *systematically* all of them may be identified from the generative power of the underlying order. Methodologically, then, it is proper to *separate* the investigations of material and other kinds of possibilities; and so to confine the latter to (for example) the various social sciences, in which the conditions, interests served and other consequences of realizing material possibilities are investigated. The same natural science suffices for all value-outlooks – for, regardless of value-outlook, all of the possibilities of interest are, or supervene on, material possibilities.

This answer leaves it open that particular strategies of the social sciences might be linked with particular value-outlooks, but not those of natural science. Natural science, according to the continued story, is *neutral*: that is, the projects of virtually any value-outlook (for example, of corporation or of movement of poor farmers) can make use of some applications made available by science in ways that strengthen or further its expression; and (in principle) the applications made available by science can serve all (currently contested) value-outlooks in an even-handed way.[30]

I take it to be uncontroversial that a considerable body of scientific knowledge gained under MS (molecular chemistry, viral and bacterial causes of disease, soil nutrients, the components of a nutritious diet, electromagnetic radiation – to give a sample) is available to be applied in ways that can strengthen the social expression of virtually any value-outlook that is actually entertained today. This explains why it is widely valued (across value-outlooks) that scientific knowledge has been gained under MS, and it provides a reason for the esteemed place that research under MS has throughout the scientific community. It does not follow that research conducted exclusively under MS (or that all research conducted under it) is valued, as distinct from inquiry in which research under MS is balanced by (or subordinated to) research conducted under alternative strategies. That is because, in contradiction with our story, the products of research under MS are in fact not neutral; the 'even-handedness' condition is not satisfied. Overall, and especially in fields like agrobiotechnology in which research is dominated by specific versions of MS, their applications favor those value-outlooks whose central practices and projects are conducted so as to further the expression of a *distinctive way of valuing control of natural objects and phenomena*. I call this distinctive way 'the modern valuation of control' (MVC).[31] MVC concerns the scope of control, its centrality in daily life, its relative unsubordination to other moral and social values, and the deep sense that control is the characteristic human stance towards natural objects; so that the expansion of technologies (informed by knowledge gained under MS) into more and more spheres of life and into becoming the means for solving more and more problems is highly valued.

Favoring value-outlooks that contain MVC violates even-handedness because the value-outlooks of various contemporary movements and groups contest MVC; and theories, consolidated under MS, do not apply *to* key phenomena and *in* significant parts of projects of importance to them. Feminist, environmental and anti-'globalization' groups bear value-outlooks that contest MVC, and also (of special interest for my argument) grassroots organizations in Latin America who adopt alternative value-outlooks that emphasize such values as '*local empowerment*,' full recognition of the entire body of human rights specified in international documents, and environmental sustainability.[32] In the agricultural projects of the grassroots organizations, phenomena of sustainable productivity, preservation of biodiversity and meeting the food and nutrition needs of the local community are of central importance, and their practices aim to preserve and enhance productive and sustainable agroecosystems over the long haul.[33] Theories developed under MS have important applications *to* these phenomena and *in* these practices, but they are limited (or subordinated); for example, they have supplied knowledge of some of the constituents and mechanisms of agroecosystems (micro-organisms, chemical nutrients), but they shed little light on the possibilities of enhancing agroecosystems – in contrast, for example, to that they shed on relations between crop yields and chemical inputs to production, and on the possibilities of production with transgenic seeds.

How can that be? Do not agroecological possibilities (like all social/ecological possibilities) supervene upon material possibilities? Despite the way our story continues, I am aware of no compelling argument that they do. Even if they do, however, it does not follow that they supervene upon material possibilities that (even in principle) may be identified under the kinds of MS (with their accompanying lexicons) that it is within human powers to develop. Some agroecological possibilities may supervene on material possibilities of such complexity, subtlety and variability with locale that human beings may not be able to identify them in the course of research conducted under MS. Be that as it may, numerous material possibilities certainly evade the grasp of theories currently or foreseeably accepted under MS. These theories also are unable to identify the agroecological possibilities, whose realization is sought in projects expressive of the values of 'local empowerment.' (This is a symptom of the lack of *neutrality* of the products of MS overall.) If these agroecological possibilities do supervene upon material possibilities, and if the latter are to be identified, then (at least for the time being) it will be *qua* agroecological possibilities, and not *qua* generable from the underlying order. It will be *qua* possibilities that things have in virtue of being part of a more or less self-regulating system, in virtue of relations and interactions they have with other constituents of the system and the role they play in its self-regulation. In so far as such agroecological possibilities may be pertinent to desired applications, identification of them will have to be gained through research conducted under alternative strategies – agroecological strategies (AES).

It is now clear why the products of research under MS can not meet the even-handedness condition. Applying current knowledge gained under MS (for example, in agrobiotechnology) may require conditions that would undermine the valued agroecological systems. It remains that knowledge gained under MS is genuine knowledge; it is expressed in soundly accepted theories or with the aid of their categories. When alternative strategies are adopted, and their results applied, that remains untouched. *Consistency* with soundly accepted theories is a mark of the rational; *applying* them need not be. Legitimation of applications involves not only that the theory has been soundly accepted, but also that its applications serve the interests of the 'right' value-outlook. When a theory is applicable only in a context where certain values are expressed and embodied, to appeal to its sound acceptance as sufficient legitimation for application implies improperly limiting the range of values that may be (rationally) held. When we separate the investigations of material and other possibilities, we study things in abstraction from the conditions for the realization of their possibilities; so it will not be part of the 'technical' investigation to figure out the social conditions under which the possibilities may be realized – so we may miss that to interact with a thing so as to realize certain of its material possibilities may actually be also to treat it as a certain type of social object.

3.2 Agroecological Strategies

AES and agrobiotechnological strategies (BTS) compete.[34] Their established theories are not inconsistent, but they encapsulate largely different classes of possibilities, which (for the most part) cannot be co-realized in the same fields. The competition concerns which class of possibilities to attempt to realize in agricultural practices: those of biotechnology which are of special interest where MVC reigns, or those of agroecology whose interest derives (in the first instance) from the values of 'local empowerment'? 'Technical scientific' issues pervade the competition: What is possible? What are the risks of application? Can the risks be suitably managed? But, provided that both BTS and AES are fruitful, the conflict is waged in the realms of values, politics, economics and so on – and where one stands in face of this conflict feeds back into the strategies one adopts in research.

Agroecologists clearly recognize this; biotechnologists often do not.[35] Sometimes it is said that the 'theory' of agroecology consists simply of ideological critique, or at best of a patchwork of opportunistically gleamed fragments of traditional local knowledge – so that the proponents of agroecology are said not to be proposing a 'scientific' research program but instead to be submitting scientific claims to ideological critique. This criticism of agroecology ignores that research under AES has been fruitful (see below), and that in fact it itself is 'ideological' rather than 'scientific.' Research under both BTS and AES gains understanding of phenomena of the world and their possibilities – and aims to do so, as well as to gain understanding pertinent to value-laden interests in application: '… political determinants enter at the point when *basic* [my italics] scientific questions are asked and not only at the time when technologies are delivered to society'.[36] If they do not appear to do so 'equally', that may be because inequalities of available material and social conditions enable research under BTS to proceed routinely without its legitimation constantly being called into question.[37] The reasons both for and against the adoption of AES, and conversely for the exclusive adoption of variants of MS, include integrally appeal to value-outlooks. The strategies are *equally* 'scientific': held to fruitfulness, and adopted (in part) because of their relations with value-outlooks. There is *not* the asymmetry that critics of agroecology sometimes claim: under MS, investigation is scientific and non-ideological, whereas research under AES is non-scientific and ideological.[38]

Those who adopt BTS, misled by the myth of *neutrality*, tend not only to downplay the empirical achievements of agroecology and to portray it as simply an ideology without link with fruitful strategies, but also to be unaware that the links of biotechnology with MVC refute the *neutrality* they claim for their own research. For them, BTS are simply particular instances of materialist strategies that enable us to identify the possibilities of things (for example, seeds) that are made available principally from using (for example) methods of genetic modification. BTS are indeed that; they are *also* those strategies whose products do and are expected to inform a particular form of technology, that is widely and almost entirely applied in practices that express highly MVC. The first description of BTS shapes research practices; the second serves to rationalize adopting them rather than other strategies.

Similarly, AES have two descriptions: first, as particular instances of general ecological strategies – that frame research on the relations and interactions between an organism and its environment, considered as a more or less self-regulating 'whole' of which the organism is an integral part – that enable us to identify the possibilities that things (seeds) have in virtue of their place in agroecological systems; second, as those strategies that are intended to provide knowledge that can inform the agricultural projects expressive of the values (for example) of 'local empowerment.'[39]

It is in virtue of the first description that AES shape research practices. I will elaborate a little. Under AES, research aims to confirm generalizations concerning the tendencies, capacities and functioning of agroecosystems, their constituents, and relations and interactions among them. These include generalizations in which (for example) 'mineral cycles, energy transformations, biological processes and socioeconomic relationships' are considered in relationship to the whole system; generalizations concerned not with 'maximizing production of a particular system, but rather with optimizing the agroecosystem as a whole' and so with 'complex interactions among and between people, crops, soil and livestock.'[40] To illustrate:

> low pest potentials [are likely] in agroecosystems that exhibit the following characteristics: high crop density through mixing crops in time and space; discontinuity of monocultures in time through rotations, use of short maturing varieties, use of crop-free or preferred host-free periods …; small, scattered fields creating a structural mosaic of adjoining crops and uncultivated land which potentially provides shelter and alternative food for natural enemies …; farms with a dominant perennial crop component …; high crop densities or the presence of tolerable levels of specific weed species; high genetic density resulting from the use of variety mixtures or crop multilines.[41]

And:

> Restoration of natural controls in agroecosystems through vegetation management not only regulates pests, but also helps to conserve energy, improves soil fertility, minimizes risks, and reduces dependence on external resources.[42]

Of particular salience are generalizations that help to identify the possibilities for productivity and sustainability of agroecosystems, where 'sustainability' has been defined in terms of four inter-connected characteristics: *productive capacity*: 'Maintenance of the productive capacity of the ecosystem'; *ecological integrity*: 'Preservation of the natural resource base and functional biodiversity'; *social health*: 'Social organization and reduction of poverty'; *cultural identity*: 'Empowerment of local communities, maintenance of tradition, and popular participation in the development process'.[43]

Theories, under AES, may be considered to be constrained so as to be able to represent sets of generalizations of the above kinds,[44] and the hypotheses (drawn from general ecological theory) that are entertained for their explanation and determining the limits of their application. The generalizations of agroecology tend to express probalilistic relations or tendencies, and they may have greater or less specificity. So, discerning the limits of application of these generalizations is especially important. Note how the generalization '… enhancement of biodiversity in traditional agroecological systems [in Latin America] represents a strategy that ensures diverse diets and income sources, stable production, minimum risk, intensive production with limited resources, and maximum returns under low levels of technology' is later qualified by '… we have still not been able to develop a predictive theory that enables us to determine what specific elements of biodiversity should be retained, added, or eliminated to enhance natural pest control.'[45]

Data are selected and sought out in virtue of their relevance for appraising these theories and for enabling phenomena, relevant in the light of the values of 'local empowerment,' to be brought within the compass of a theory's applicability. Obtaining the data often requires subtle, regular, painstaking, accurate observation and monitoring of a multiplicity and heterogeneity of details in the agroecosystems. The skills for this are usually only developed by local farmers themselves, so that gaining the data depends on the collaboration of local farmers and the utilization of their experience and knowledge, and the lexicon in which they are reported will reflect the distinctions and categories of this experience. Agroecology cannot be pursued with a sharp distinction between the researcher and the farmer; the farmer's observations are essential to the conduct of the research. Quantitative data are often pertinent: counting the number of pests in a given area, measuring the size of crop yields, amount of water available and so on; statistical comparisons (for example) of pest populations across (for example) monocultures and polycultures, or of the yields of different crops when different methods are used. Experimental data are sought both to support statistical comparisons, *and* to demonstrate that possibilities can be realized in agroecosystems with certain characteristics, for example: '… it is possible to stabilize the insect communities of agroecosystems by designing and constructing vegetational architectures that support populations of natural enemies or have direct deterrent effects on pest herbivores.'[46] In agroecological contexts, an 'experiment' involves introducing, for the sake of observing its systemic effects, a modification (under an investigator's control) of an agroecological system. Given the local distinctiveness of agroecosystems, the mark of a 'good' experiment cannot be its replicability across diverse environmental and social conditions. Note that control is involved in agroecological experiments and farming practices, but subordinated to the values of 'local empowerment.'

Relevant data are often obtained from the study of farming systems in which *traditional methods informed by traditional local knowledge* are used. These systems are appropriately submitted to empirical scrutiny because agroecological studies have shown 'that traditional farming systems are often based on deep ecological rationales and in many cases exhibit a number of desirable features of socioeconomic stability, biological resilience and productivity.'[47] They exemplify many known agroecological principles and others can be expected to be extracted in the course of studying them.[48] They can (with adaptations suggested by research findings) be enhanced with respect to all four of the characteristics listed above, and especially with respect to 'cultural identity' they are often uniquely appropriate for the activities of poor, small farmers. It is worth noting that the methods used in these systems have been tested rigorously in practice, and have been particularly effective (reflecting the experimental approach of traditional farmers) over the centuries in 'selecting seed varieties for specific environments'[49] – these are often the seed varieties (or the original sources of them) that are modified genetically in biotechnology research and practice.[50]

3.3 Adopting a Strategy and the Social Location of Scientific Research

At least in some fields, there can be multiple strategies that compete in the way described above; and each of the competing strategies may be fruitful. Then there arises the question of which strategy to adopt, one for which different answers may be proposed and acted on by different investigators. I have suggested that actual answers (explicitly or implicitly) draw upon mutually reinforcing relations between adopting strategies and the value-outlooks whose interests would be served especially well by applications of knowledge gained under the strategies.

There is, however, a general reason to opt in favor of developing research under *some* strategies other than MS: to test whether all possibilities, or all material possibilities upon which non-material possibilities have been assumed to supervene, can become grasped under MS. By identifying possibilities that are not identical with possibilities currently encapsulated by soundly accepted theories under MS, we can pose concrete challenges for research under MS to meet. This reason would not appeal to Kuhn; he holds that such challenges are unnecessary since, in due course, anomalies will accumulate in the normal unfolding of MS. However, there may be bounds to MS, while within the bounds there remain unlimited possibilities to be identified. Only tests of the kind indicated here can hope to identify these bounds. (I am not sure that Kuhn recognized this.) By identifying possibilities of the kinds indicated, it can be probed empirically whether or not there are bounds to the development of MS. This general reason sits in tension with the rationales for adopting specific strategies, especially when we remember that there is also competition for resources for conducting research. Resources devoted to probing the limits of MS in this way would be resources taken away from pursuing more favored projects and *de facto* giving the resources to support research whose strategies gain their primary rationale from competing value-outlooks. Only a satisfactory resolution of this tension, I believe, could restore *neutrality as* a compelling value of scientific practice.[51] The tension is heightened when we consider the legitimation of applications.

Kuhn's picture, recall, portrays applications mainly as consequences of scientific developments; and developments under MS have identified numerous material possibilities that have become, and are continuing to become, realized in applications at an increasingly rapid rate. The *efficacy* of applications depends on the input of sound scientific knowledge that can be provided (for many applications) by research under MS. The *legitimation* of some applications depends also on claims about the possibilities of things. Consider: under MS, means (involving developments of biotechnology) may be identified for producing food sufficient in quantity to continue to feed the world's population. Applying the knowledge thereby obtained is legitimated, however, only if there are no 'better' ways of producing sufficient amounts of food – ways, for example, that would be part of agroecosystems that were structured so that the food is not only produced, but also so that everyone is actually fed sufficiently and nutritiously, and that sustainable (and improving) and productive agroecosystems are maintained.[52] But the possibility of producing sufficient food by *developed and expanded* uses of agroecological methods cannot be investigated under MS. So research under MS cannot provide a crucial item of knowledge (or the means for attempting to gain it) needed to legitimate endorsing biotechnological methods as essential to the solution of the world's food problems.

The proponents of bringing biotechnological methods to the core of agricultural production respond that there is *no* evidence that developed agroecological methods could produce sufficient food. In responding to this assertion it is important to keep in mind that producing sufficient quantities of food to feed everyone does not imply that everyone will be fed. Currently, sufficient food is produced, but hunger persists.[53] Given that agrobiotechnology plays an integral role in the global economic system, under which hunger currently persists, one might wonder why the expansion of production of food promised by the new methods will be any more likely to lead to the hungry being fed. Who is fed, and who is not, is not independent of the methods of production. Even if there were strong evidence that enhanced agroecological methods could not produce sufficient food to feed everyone, it might still be the case that agroecology needs to be developed so that the currently hungry and their descendants will be fed. The legitimation of the furthering of agroecology needs only this more modest claim which, I believe, is the claim that agroecologists actually make. Perhaps, in order that everyone be fed, a variety of farming methods will have to be used. The issue is an empirical one, but investigations conducted exclusively under MS cannot adequately address it. It can only be responsibly investigated within a theoretical framework that investigates the full causal nexus of production and consumption of food, and more generally of human well-being, and in a process that is responsive to the needs, interests and value-outlooks of everyone.

As things stand, it is true that the evidence is *less than compelling* that agroecological methods can be enhanced and expanded to produce sufficient food to feed everyone. However, that could be because, while AES have displayed a measure of fruitfulness, their limits have effectively been reached; or because, due to lack of the necessary social conditions and material resources, there has been much less research conducted under AES than under BTS. This matter could be explored empirically by providing conditions to further develop agroecology in those areas where there is hunger and an available rural workforce (thus furthering it under the legitimation of the modest claim referred to in the previous paragraph, in areas where its effectiveness has been repeatedly demonstrated;[54] this would enable virtually risk-free investigation of the possibilities of agroecological production.

The proponents of biotechnological methods see little urgency in conducting such an investigation because, I think, the widespread implementation of agroecological methods would be incompatible with the social structures, values and policies under which biotechnology is developing. For them, agroecological methods cannot produce sufficient food because they cannot be developed under these social conditions. There is, for these proponents, no *better* way to produce the needed food, because biotechnological methods are confirmed as providing the most efficacious of the available possibilities whose realization could be informed by theories established under MS, and thus could most usefully further the expression of MVC; and, for them, furthering MVC (and so, at the present moment, fitting into the neo-liberal global economic project) has become a condition on a legitimated way. In short, the legitimation available to be offered of prioritizing biotechnological methods – without begging questions whose empirically grounded answers require developments of AES – does not rest upon empirical confirmation that agroecological methods are insufficient for producing the food. It rests upon commitment to MVC (and the values of 'globalization'), upon valuing the material possibilities of biotechnology because to realize them is, at one and the same time, to realize valued social possibilities.

There are no 'scientific' reasons to decline to appraise empirically the possibilities of agroecology. Given that applications involve issues not only of efficacy, but also of legitimation or social value, it is just arbitrary to insist that what counts as a 'scientifically' interesting phenomenon is determined only in view of the internal unfolding of the scientific tradition and not also by interests connected with application. Thus, the competition between BTS and AES cannot be dissolved by appealing to the general character of science. The marginalization of AES in the mainstream, I have suggested, is explained (when we probe for the reasons) not because, after adequately providing for efforts to develop them, serious doubts about their fruitfulness have been confirmed. Rather it is because they cannot lead to applications of interest for MVC; and perhaps also because, if their fruitfulness were confirmed, the legitimation of prioritizing biotechnology in agriculture would be challenged – though in fact the proponents of biotechnology tend not even to entertain that the far-reaching fruitfulness of AES might, given the opportunity, be confirmed.[55] Conversely, the reasons for adopting AES (which, I repeat, draw upon basic knowledge gained under MS in all sorts of ways) as an alternative to BTS are connected with critique of MVC and with holding such competing value-outlooks as that of 'local empowerment.' Either way, adoption of strategies, and thus the character of research conducted, is unintelligible if separated from the social location of scientific practices and their applications; and thus, in turn, social location can serve as a ground (but not one that downplays the importance of fruitfulness) for critique of scientific practices, and as a source and condition of alternatives.

4 Conclusion

The object of scientific inquiry is phenomena as grasped under a strategy, so much so that it varies with strategies and cannot be characterized in strategy-neutral terms. That is Kuhn's insight. Strategies, a key component of scientific methodology, are historically variable, and so too is the object of scientific inquiry.

To understand phenomena is to describe and explain them, *and* to identify the possibilities they admit. A strategy has the resources to identify a particular class of possibilities. Any one strategy is worthy of adoption only if, given the opportunity and appropriate resources, it shows itself to be fruitful: that is, if it is successful in actually identifying possibilities of the relevant class (encapsulating them in soundly accepted theories). Competing strategies explore classes of possibilities (often of the same phenomena, for example, seeds) that cannot be co-realized: for example, realizing (to any significant extent) the possibilities of transgenic seeds cannot be co-realized with certain possibilities of productive and sustainable agroecosystems.

When two fruitful strategies compete, what are the reasons to adopt one of them rather than the other? Since both are fruitful, reasons based exclusively on cognitive (epistemic) value cannot favor one rather than the other. As illustrated in the discussion of Section 3, my answer is: Adopt the one that enables us to gain understanding that is applicable *to* phenomena and (where appropriate) *in* practical projects of significance for our value-outlooks, thus the one that identifies possibilities that, if realized on application, would further these projects. This provides a good reason to adopt a strategy without, at the same time, denying that the scope and value of the basic understanding gained in scientific research transcend interest in applications. It is a reason that points to the (social) value of research conducted under the strategy.

Both fruitfulness and applicability are necessary conditions for the adoption of a strategy. We adopt a strategy partly for the sake of gaining theories that are applicable in ways that are significant for our value-outlooks. In a particular field of research, there may be no relevant disagreements across value-outlooks about what are the phenomena and projects for which applications of theories are desired. Then competing strategies are unlikely to emerge. Different value-outlooks, however, may (in some fields) lead to different appraisals of the significance (social value) of applications, and thus to their respective adherents adopting competing strategies (for example, AES and BTS). Where this happens, a case can be made that a plurality of strategies should appropriately be supported within the whole scientific community (despite the resulting tensions that would be occasioned by the fact that the classes of possibilities being explored are not co-realizable in the same contexts). Moreover, if my suggestion is correct, it will be no surprise that one kind of strategy comes to be adopted virtually exclusively in the scientific community, and that adopting a strategy is not generally recognized as a matter of choice or as in need of rational support, when in it and its supporting institutions there is hegemony of values (for example, MVC or those of the global economy).

According to the picture I have offered, application plays a central role in shaping scientific practice. It is not just a consequence (or instrumental partner) of successful research, but where it is valued in social practices that one endorses, it is part of the very reason to adopt a strategy. We might put it: Possibilities, in so far as they are identical to possibilities for application, *partly* constitute the object of scientific inquiry. Such possibilities, of course, are objects of social value and historically and culturally variable, functions of the social location of the scientific practices. The strategies we adopt are those suitable for exploring these possibilities, and so they too must vary as a function of the social location of scientific practices. Thus, applications – successful, desired, anticipated – feed back so as to influence at the most fundamental methodological level the way in which scientific investigation is conducted.

Notes

1 I wish to thank the National Science Foundation (SES–9905945) for partial support of the writing of this paper, and Richard Eldridge and the editors for helpful comments.
2 Margolis (1995), p. 321.
3 At the present moment this supposition seems to be rapidly losing its grip. Especially in the biological sciences (for example, concerning the Human Genome Project), the components of the underlying order are increasingly becoming thought of as objects to which one may hope to obtain intellectual property rights; thus, among other things, as historically constituted property and commodities.
4 See Lacey (1999a), Ch. 6.
5 Ibid., pp. 124–6.
6 Supposition 3 will play no role in the present argument. Properly articulated, using the notion of cognitive values, it may be compatible with the denial of supposition 4 – see my defense that theories, regardless of the strategies under which they are developed, may become accepted in accordance with *impartiality* (ibid., Ch. 10).
7 Ibid., pp. 13–14.
8 Cognitive values are the criteria of the cognitive or epistemic value (rational acceptability) of a theory. I have discussed them extensively elsewhere (ibid., Ch. 3), and defended that they can be distinguished from social, moral and other kinds of values (Lacey, 1999b). I will not address how the history of the cognitive values might bear upon the historicity of science; this may have implications regarding supposition 3 that I will not entertain here.
9 Lacey (1999a), Ch. 6.
10 Ibid., Ch. 7.
11 Kuhn (1970). For the history of Kuhn's use of 'paradigm' and the terms he introduced in subsequent work to refine and replace it ('disciplinary matrix', 'structured lexicon'), see Hoyningen-Huene (1993) and Sankey (1994). I have elaborated the notion of strategy (my terminology, not Kuhn's), as well as ideas introduced in the next few paragraphs, in detail elsewhere (Lacey, 1999a; 1999b).
12 Under certain conditions, that are connected with applications (see next section), theory choice may be made across strategies (Lacey, 1999a, Chs 7, 10; 1999e).
13 Lacey (1999a), pp. 62–6.
14 I have emphasized elsewhere (ibid.) the variety to be found among materialist strategies. Not all of them represent reductionist tendencies, and not all of them require that laws be deterministic. I say that MS are adopted 'virtually exclusively' in modern science:

ecology (which is not confined to deployment of MS) usually is considered to fall under the umbrella of modern science.

15 Kuhn (1970) has developed this idea in a rich way using the phenomenological notion of 'world.' The objects of the 'world' in which investigations under MS are conducted are partly constituted by the practices themselves. I have discussed this in Lacey (1999a, Ch. 7; see also Lacey, 1999c; 1999e).

16 Lacey (1999a), Ch. 5.

17 I will not keep repeating the qualification: 'in the relevant field'. It applies to all the remarks about strategies (and research framed by a strategy) that follow in this chapter.

18 That helps to explain why often it is thought that science *is just* that inquiry conducted under the currently dominant strategies. (Modern science *is* inquiry conducted under MS.) Since normally there is no controversy about MS in the scientific community, their role can easily remain hidden so that it is not recognized that there may be other strategies, and investigation that might be being conducted under another strategy tends to be dismissed as 'unscientific.' This theme comes up again in Section 3: *agroecological strategies* (see also note 55).

19 Cf. Hoyningen-Huene (1993), pp. 241–3.

20 Lacey (1999a), pp. 172–5; 1999e.

21 Kuhn (1956).

22 Kuhn (1970).

23 Lacey (1999a), Ch. 7.

24 Cartwright (1999).

25 Of course, this does not prove that there is no such underlying order, as Sankey (1997) – see Lacey (1999c) – has emphasized.

26 Lacey (1999a), 14–15.

27 Here I only present my alternative picture, showing how it illuminates an important contemporary controversy. See Lacey (1999a, Ch. 7; 1999e) for fuller argument.

28 Endorsing the values expressed and furthered by the prevailing social order *may* provide a good *reason* to adopt MS virtually exclusively – of course, the reason is only as good as the grounds for endorsing these values (Lacey (1999a), Ch. 6). That they are widely endorsed, and expressed deeply in dominant modern economic and political projects, may largely *explain* that scientific research is conducted almost exclusively under MS.

29 Kloppenburg (1987); Lacey (1999a, Ch. 8) and the references there.

30 Lacey (1999a), Ch. 4; 1999b.

31 See Lacey (1999a), pp. 111–30, for a detailed analysis of MVC and for the argument – also Lacey, 1999d.

32 Lacey (1997); Lacey (1999a), Ch. 8.

33 Control of natural phenomena is, of course, a value for them – as it is in every culture – but, unlike in MVC, it is subordinated to the listed core values.

34 My account of *agroecology* here is derived from the numerous writings of Altieri (especially Altieri, 1995) with some adaptations of terminology (that involve little strain) so as to fit into my general analytic framework. (See also Lacey, 1999a: Ch. 8.)

On *biotechnology*: '... in essence [biotechnology] implies the use of microbial, animal or plant cells or enzymes to synthesize, breakdown or transform materials. ... Traditional biotechnology refers to the conventional techniques that have been used for many centuries to produce beer, wine, cheese and many other foods, while "new" biotechnology embraces all methods of genetic modification by recombinant DNA and cell fusion techniques, together with modern developments of "traditional" biotechnological processes' (Smith, 1996: 2–3). In the text I am using 'biotechnology' in the sense of 'the "new" biotechnology.'

I am also taking it to refer to a field of scientific research: that conducted under BTS, research that aims to produce knowledge that can enhance the methods specified in the quote. Thus, depending on contest, 'biotechnology' may refer either to a field of scientific research or to specific methods deployed in agricultural practices.

Note that the way in which AES and BTS compete does not preclude that each may draw from the positive results of the other *in limited respects*. In this chapter I will not explore whether and how this might happen. On the relationship of the kind of competition, exemplified by that between AES and BTS, with what Kuhn has called 'incommensurability,' see Lacey, 1999e; and see the next paragraph for how it involves the difficulties of communication that Kuhn diagnoses to be part of incommensurability.

35 See the exchange between Altieri and Rosset (2000) and McGloughlin (2000).
36 Altieri (1994), pp. 150–51.
37 See note 18 above.
38 McGloughlin (2000).
39 In both cases (BTS and AES) the adopting of the strategies is rationalized (in part) by reference to particular values. This does not *per se* challenge the *impartiality* (sound acceptance) of the results consolidated under either strategy; it may their *neutrality*. In the case of AES, since objects (including agroecosystems themselves) are not abstracted from their places in human experience and social relations, values enter into the subject-matter of the investigation: under what conditions are certain values (for example, social justice, cultural identity) able to be further embodied? (Under MS, all comparable questions are pushed into the social science inquiries that may inform applications.) Note that the questions (posed under AES) are about the degree of embodiment and manifestation of the values; reaching empirically based results about them (as distinct, perhaps, from having an interest in them) is logically independent of endorsing the values. There can be *impartial* results about the degree of manifestation and embodiment of values (Lacey, 1999a, Ch. 2).
40 Altieri (1987), pp. xiv–xv.
41 Altieri (1999), pp. 24–5.
42 Altieri (1994), p. 150.
43 Altieri et al. (1996), pp. 367–8.
44 See Lacey (1999a), pp. 193–6 for further discussion.
45 Altieri (1994), pp. 7, 38.
46 Ibid., p. 7.
47 Altieri (1987), p. xiii; for details and examples, see Altieri, 1995: Ch. 6.
48 Altieri (1995), p. 143.
49 Ibid., p. 116.
50 Kloppenburg (1987); Lacey (1999a), Ch. 8.
51 Lacey (1999a), Ch. 10.
52 Altieri and Rosset (2000); Kloppenburg and Burrows (1996). Other important issues are also involved in the legitimation of prioritizing (or even using) biotechnological methods: for example, concerning possible undesirable health and environmental side-effects. They have been widely discussed (see, for example, Rissler and Mellon, 1996).
53 Boucher (1999).
54 Altieri et al. (1996).
55 My explanation is consistent with it being the case that, in the consciousness of researchers in biotechnology, they are simply following through on the latest options provided under MS, with no issue of choice of strategy involved (see note 18). If there is no choice of strategy, then criticism of biotechnology becomes seen simply as criticism of engaging

in research *per se*, or as having the effect of threatening funds for research. Some recent reactions of biotechnology researchers to criticism have been of this kind. Ironically, they see threats to the 'autonomy' of science coming more from their critics than from the corporate sponsors of much of their research.

References

Altieri, M.A. (1987), *Agroecology: The Scientific Basis of Alternative Agricultures*, Boulder, CO: Westview.

Altieri, M.A. (1994), *Biodiversity and Pest Management in Agroecosystems*, New York: The Haworth Press.

Altieri, M.A. (1995), *Agroecology: The Science of Sustainable Agriculture*, 2nd edn, Boulder, CO: Westview.

Altieri, M.A. (1999), 'The Ecological Role of Biodiversity in Agroecosystems', *Agriculture, Ecosystems and Environment*, 74, pp. 19–31.

Altieri, M.A. and Rosset, P. (2000), 'Ten Reasons Why Biotechnology will not Help the Developing World', *AgBioForum*, 2: <http://www.agbioforum.org>.

Altieri, M.A. , Yurjevic, A., Von der Weid, J.M. and Sanchez, J. (1996), 'Applying Agroecology to Improve Peasant Farming Systems in Latin America', in R. Costanza, O. Segura and J. Martinez-Alier (eds), *Getting Down to Earth: Practical Applications of Ecological Economics*, Washington, DC: Island Press.

Boucher, D.H. (ed.) (1999), *The Paradox of Plenty: Hunger in a Bountiful World*, Oakland, CA: Food First Books.

Cartwright, N. (1999), *The Dappled World: A Study of the Boundaries of Science*, Cambridge: Cambridge University Press.

Hoyningen-Huene, P. (1993), *Reconstructing Scientific Revolutions: Thomas S. Kuhn's Philosophy of Science*, Chicago: University of Chicago Press.

Kloppenburg, J., Jr (1987), 'The Plant Germplasm Controversy', *Bioscience*, 37, pp. 190–98.

Kloppenburg, J., Jr (1991), 'Social Theory and the De/reconstruction of Agricultural Science: Local Knowledge for an Alternative Agriculture', *Rural Sociology*, 56, pp. 519–48.

Kloppenburg, J., Jr and Burrows, B. (1996), 'Biotechnology to the Rescue? Twelve Reasons Why Biotechnology is Incompatible with Sustainable Agriculture', *The Ecologist*, 26, pp. 61–7.

Kuhn, T.S. (1956), *The Copernican Revolution*, Cambridge, MA: Harvard University Press.

Kuhn, T.S. (1970), *The Structure of Scientific Revolutions*, 2nd edn, Chicago: University of Chicago Press.

Lacey, H. (1997), 'Neutrality in the Social Sciences', *Journal for the Theory of Social Behavior*, 27, pp. 213–41.

Lacey, H. (1999a), *Is Science Value Free? Values and Scientific Understanding*, London: Routledge.

Lacey, H. (1999b), 'On Cognitive and Social Values', *Science and Education*, 8, pp. 89–103.

Lacey, H. (1999c), 'Philosophically Reconstructing Kuhn', *Metascience*, 8, pp. 188–92.

Lacey, H. (1999d), 'The Ways in which the Sciences are and are not Value Free', in P. Gardenfors, K. Kijania-Placek and J. Wolenski (eds), *Proceedings of the 11th International Congress of Logic, Methodology and Philosophy of Science*, Routledge: London.

Lacey, H. (1999e), 'Incommensurability and "Multicultural Science"', in P. Hoyningen-Huene and H. Sankey (eds), *Incommensurability and Related Matters*, Routledge: London.

Margolis, J. (1995), *Historied Thought, Constructed World: A Conceptual Primer for the Turn of the Millennium*, Berkeley, CA: University of California Press.

McGloughlin, M. (2000), 'Ten Reasons Why Biotechnology will be Important to the Developing World', *AgBioForum*, 2: <http://www.agbioforum.org>.

Rissler, J. and Mellon, M. (1996), *The Ecological Risks of Engineered Crops*, Cambridge, MA: MIT Press.

Sankey, H. (1994), *The Incommensurability Thesis*, Aldershot: Avebury.

Sankey, H. (1997), *Rationality, Relativism and Incommensurability*, Aldershot: Ashgate.

Smith, J.E. (1996), *Biotechnology*, 3rd edn, Cambridge: Cambridge University Press.

Chapter 7

Kuhn, Different Worlds and Science as Historical

Tom Rockmore

The prevailing idea that claims for truth and knowledge are ahistorical has a number of presuppositions which are initially formulated in Platonism, and re-stated much later in contemporary realism. These include the conviction that there is a single, permanent, mind-independent external world, and that we know it as it is, for instance in isolating items of knowledge, or so-called facts which are strictly independent of frames of reference, or at that we at least know the world as it is from the different angles of vision provided, say, by the difference between folk and scientific perspectives, or between different scientific theories.

The debate about knowledge in the last century successfully refuted, or at least weakened, all the presuppositions of the traditional ahistorical approach to epistemology, whose negation leads to a historical conception of truth and knowledge. Those committed to an ahistorical conception of knowledge frequently defend forms of realism, whose defenders are often discouraged, increasingly beleaguered.

The analytic critique of classical empiricism undermined confidence in the belief classical empiricism shares with Descartes that the mind can come into direct, immediate contact with the world, leading to suggestions in the later Wittgenstein and others that claims to know depend on their relation to a conceptually prior context in which they are embedded and within which they are meaningful. Yet many writers, even those who participated in the analytic attack on classical empiricism, have been reluctant to accept its consequences, unwilling to give up the cognitive advantage to be gained in being able to claim to know the world as it is. In reaction against the relativity of contextualism, the old view that we directly know the mind-external independent world as it is has been reasserted by Davidson and Putnam. Davidson has re-committed, and then recently abandoned, a correspondence theory of truth, and Putnam has abandoned internal realism for his new, quasi-Jamesian form of naive realism.

The aim in both cases seems to be to limit the damage done by the analytic dismantling of classical empiricism by avoiding the radical, anti-realist, skeptical consequences drawn by Rorty and even the contextualist maneuvers introduced by Wittgenstein. The prize to be gained is to skirt the difficulties of renewed skepticism in maintaining a claim to cognitive objectivity through a form of realism even while abandoning classical empiricism. The idea is once again that there is a mind-independent external world which we do know when we know. Both for Plato and for recent realists, the presupposition of this view is that the world is of a certain type, a view which is sometimes expressed as metaphysical realism, or the assertion that 'the world consists of some fixed totality of mind-independent objects.'[1]

The view that the world is of a certain type is independent of ways it can be known. Writers such as Putnam, Davidson, T. Nagel, B. Williams and Nozick differ in their precise approach to knowledge, but not in their conviction that we find out things about the way the world is. Rorty denies this. Although Rorty the skeptic and the others differ about the claims to be made about the world, they do not differ at all about the realist conviction going all the way back to Plato that there is at least something out there to be known as it is. This assumption is common to those who think there is one way to map it, to those, more liberal, who hold it can be explored in different ways, and finally to those who, dispairing of knowledge, think it cannot be explored or even known at all. Partisans of these and other approaches to knowledge disagree on many other things, but agree at least that, as Bishop Butler famously thinks, everything just is what it is and not another thing.

But what if this conviction underlying realism, and underlying Platonism as well, which is its conceptual origin, could be challenged? What if there were reason to think that the assumption that there is, as Putnam puts it, a single world consisting of some fixed totality of mind-independent objects were a merely gratuitous assumption and we could get along without it? What would change if this argument could be made out?

Certainly, one thing which would change would be the very idea that to know is to know something permanent, which, hence, can be known in a way lying beyond time and space, hence ahistorically. If there were no permanent real world, if the world itself were impermanent, say, in some way a historical variable, a 'construct' or 'product' of the knower, then the problem of objective cognition would have to be understood differently than it has been, and it would need to be admitted that it might centrally depend in some way or other on history, even that knowledge itself is intrinsically historical.

Now this would amount to nothing less than a sea-change in the way truth and knowledge have been understood in the Western tradition. In the West, since Parmenides, the first thinker who raises the problem of knowledge in a recognizably modern fashion, on the basis of claims to know the mind-independent world as it is, writers from Parmenides to the present routinely understand knowledge as truth, or an ahistorical grasp of what is in independence of the knower. This normative view of knowledge as truth beyond history, only inessentially related to time and place, turns into a view of knowledge as intrinsically historical, that is, indexed to the historical moment, if it turns out that not only our claims to know but even the cognitive objects themselves are artifacts of human history.

1 Kuhn and the Historical Approach to Knowledge and its Object

In this respect, Kuhn offers an interesting case: he denies that we can pick out facts which are not frame-relative; and, consistent with this view, he further denies that we can know mind-independent external reality as it is in favor of a conception of science in which there are different paradigms of a world we can never know that we know as it is, and in which even what is meant by 'world' is dependent on the theory or theories through which it is known.

In his denial of independent facts, Kuhn is very close to holists like Duhem and Quine as well as to contextualists of various kinds. There is a close link between epistemological holism, or the view that, as Quine succinctly puts it, theories meet their fate as a whole, and contextualism, according to which both facts and theories cannot be isolated from the larger context in which they arise, for which particular theories are meaningful, and in relation to which they are justified.

Orthodox Christianity offers an early version of contextualism in a form which few for other than theological reasons would now defend. Over the centuries the Church evaluated theories and even observations as true or false as a function of their agreement or disagreement with its own view, which, on the basis of revelation, was taken as true and immutable. The rejection of Galileo's efforts to support Copernican astronomy, including his later ecclesiastical prosecution, perhaps even the failure over centuries to admit that theology cannot legitimately impose its will on science, cannot be understood without acknowledging Christian dependence on a view of knowledge presupposing agreement with a wider framework, whose truth is never shown, and which cannot be shown, but which is presupposed as true.

Contemporary contextualists differ from their theological predecessors, Christian or otherwise, in denying that presupposed claims for truth by revelation of the wider context or contexts governing the acceptance or rejection of theories or observations. Contextualists deny that facts can be isolated from conceptual frameworks, which in turn they regard as indexed to wider contexts. These can be either very general scientific perspectives, such as the difference between modern, quantitative science and ancient, qualitative science, or a few very general theories, say Darwinian evolution, quantum mechanics and relativity.

All forms of cognition, including science, concern a relation between various theories and their cognitive objects. The latter can be understood in different ways. One is that the cognitive object is simply mind-independent, hence discovered by the knower as it is in independence of us. A second is that the cognitive object is dependent on the knower in various ways, as in Kant's critical philosophy. The most plausible interpretation of the critical philosophy is as a theory of of knowledge about a mind-independent external world which we can only know within the limitations imposed on us by the nature of the human mind. Third, the cognitive object can be regarded as a kind of 'construct' which, while dependent on empirical constraints, can take very different forms.

For a Kantian, the conceptual object is a necessary source of what, from the angle of vision imposed by the conceptual machinery of the human mind, human beings can possibly know. A very similar kind of epistemological vision is being developed at present under Cartesian auspices by Chomsky, whose grasp of Descartes is apparently precarious, but who straightforwardly contends that the ability to learn to speak different natural languages depends on a series of invariant rules which are hard-wired, so to speak.

Kant and those with similar theories are distinguished by the idea that conceptual frameworks for processing empirical information are both necessary and necessarily immutable. It is especially difficult to argue for the latter claim. If we follow Kant in denying claims for direct knowledge, then it could be inferred that all our knowledge claims are perspectival, hence infected by our relation to the surrounding context, or at least lodged within a particular theory.[2] Yet it does not follow, as Kant contends, that the frameworks themselves are immutable, or fixed. Wittgenstein, for instance, contends that claims to know are true or false with respect to a form of life which is itself neither true nor false. He is at least aware of the possibility that forms of life, hence conceptual frameworks, could change. But he never develops the consequences of this point for his theories. He does not, for instance, suggest that since claims for truth and knowledge are framework-relative, they are indexed to a time and place, to a particular historical moment, hence historical even though his particular form of contexualism points in that direction.

A different kind of problem is raised by the degree of 'stability' of the cognitive object. In this respect, Kuhn's conception of different worlds is extremely interesting. For it hints that paradigms or general theoretical perspectives englobing particular theories as well as empirical observation, or the relation to empirical constraints, are related to changes in our theories of the world, in what we mean by 'world,' or even to changes in the world itself.

These ideas are obviously very different. It is one thing to say that our views are perspectival, but the different perspectives relate to the same thing, which is unaffected by our view about it, another to say our understanding of 'world' changes, and still another to suggest that the world itself is not stable, but in some sense unfixed, variable, mutable, for instance dependent on the very conceptual framework through which it is cognized. Kant holds a form of the former view in his conception of the thing in itself which stands outside experience through which we seek to know it. Hegel features a version of the latter approach, in which what we seek to know is not independent of but rather dependent on the way we seek to know it, in which the object is itself an artifact of the theory, perspective, or conceptual framework.

Kuhn's view of different worlds seems to feature both views. Like Kant, he sometimes indicates that there is a mind-independent external world (which we seek to know) which does not change as a function of our knowledge of it. Yet he also sometimes suggests that the world itself is no more than a 'construct' of our efforts to know it. For our purposes, the latter, more radical view is also more interesting. The former view suggests that at different times we have different views of the world, say the different views which correspond to different stages in the development of the natural sciences. The latter view suggests, on the contrary, that the world itself is intrinsically historical, since it is not fixed, but varies as a function of our understanding of it. In the latter case, not only is knowledge itself historical, but the very idea of a single, immutable, mind-independent object, on which we have different perspectives, or points of view, at different times, must be abandoned in favor of the idea that the world itself changes as our ideas about it change.

2 Kuhn on Different Worlds

Kuhn's concept of different worlds, which arises in the context of his discussion of paradigm shifts, is related to the much criticized, widely misunderstood incommensurability of different paradigms, his term for a conceptual framework.[3] Since proponents of competing, incommensurable paradigms begin from different approaches, they often have difficulty in talking productively to one other, resulting in a so-called failure to make contact with each other's viewpoints. In his account of the resolution of revolutions in *The Structure of Scientific Revolutions*,[4] Kuhn gives three reasons for this putative failure: disagreement about the list of problems to be solved,[5] a misunderstanding between the two schools and their different worlds,[6] and the fact that the proponents of different paradigms belong to different worlds. In reference to the latter point, he famously writes: 'In a sense that I am unable to explicate further, the proponents of competing paradigms practice their trades in different worlds.'[7]

This passage suggests a close but unclarified relation between different worlds and different scientific paradigms. The most obvious interpretation is that representatives of different scientific perspectives belong to, in fact literally live in, different worlds, which derive from, hence can be said to exist as, a function of the different scientific theories. If that is true, then a scientific theory is related to a perspective, point of view, conceptual framework or way of looking at the world, as a result of which what is meant by 'world' and even the world itself changes. On this interpretation, the world itself is not independent of theories about it, but rather an artifact of the theories related to it. Neither Newtonian mechanics nor Einsteinian relativity is a world. But each calls into being a different world, or type of cognitve object. Newtonian mechanics points toward a world of point-masses in which planets follow elliptical orbits which can be described in terms of spatial and temporal dimension; Einsteinian relativity points toward field theories in which orbits are distortions in a single space–time.

A new paradigm, which is not a different world, is rather the condition for the emergence of the latter. As the Church correctly anticipated, hence feared, the Copernican astronomical revolution, which incorporated a new astronomical paradigm, had consequences which far surpassed mere astronomy. It brought with it a new way of seeing a great many things, a way different from and even in conflict with the Christian theological view. It gave birth to a new, different world in which theology had at most a very reduced role to play, leading to an irreversible secularism, which even the most fundamentalist movements of our time are unlikely to reverse or to hinder more than momentarily.

This way of reading Kuhn's idea of different worlds emphasizes the radical nature of an approach which not only indexes facts about the world to theories, but further indexes the world itself to the theories about it. But there is another, weaker way of reading the same idea in which, although the facts are indeed dependent on the theories about the world, the world itself is not. Part of the difficulty of Kuhn's position is making sense of precisely how he understands the relation between theory and world, hence the idea of different worlds. At different points in his writings, Kuhn seems to emphasize both both readings. At least three further passages bear Kuhn's understanding of the concept of different worlds. Taken together, these passages show that and how Kuhn continued to reflect on the idea of different worlds without ever arriving at a final view. These passages include one on different worlds from the first edition of the book in 1962, another on two main senses of 'paradigm' from the postscript written in 1969, and a third on the relation between different worlds, different data and similar stimuli in an article from 1974 which was later reprinted. Here are the passages:

> Lavoisier, we said, saw oxygen where Priestley had seen dephlogisticated air and where others had seen nothing at all. ... At the very least, as a result of discovering oxygen, Lavoisier saw nature differently. And in the absence of some recourse to that hypothetical fixed nature that he 'saw differently,' the principle of economy will urge us to say that after discovering oxygen Lavoisier worked in a different world.[8]

> [I]n much of the book, the term 'paradigm' is used in two different senses. On the one hand, it stands for the entire constellation of beliefs, values, techniques, and so on shared by the members of a given community. On the other, it denotes one sort of element in that constellation, the concrete puzzle-solutions which, employed as models or examples, can replace explicit rules as a basis for the solution of the remaining puzzles of normal science.'[9]

> In *The Structure of Scientific Revolutions*, particularly chapter 10, I repeatedly insist that members of different scientific communities live in different worlds and that scientific revolutions change the world in which the scientist works. I would now want to say that members of different communities are presented with different data by the same stimuli.[10]

These passages appear to suggest different, incompatible views of science. The first, justly famous passage describes the change in the basic underlying conceptual framework and, since he rejects the idea of what he calls a fixed nature, a change in the world itself as opposed to knowledge about it. This suggests that science is an essentially historical enterprise in that as it changes, not only its theories but even its objects, which are identified by its theories, change. The second concedes that in his book the term 'paradigm' is used to stand for a shared constellation of beliefs and so on, what is often called a *Weltanschauung*, as well as for a conceptual model belonging to that constellation, which is employed in what Kuhn calls normal science. Science here appears as responsive both to intra-scientific as well as to extra-scientific factors, hence as dependent on different types of changing contexts. The third claims there is a single mind-independent, real independent world, which is the source of the same stimuli for all observers, but which, since observers differ, presents different data to members of different communities. The world does not change, although our way of viewing it does change as a result of changes in our theories about the world. Science now takes the form of a rational enterprise, driven by intra-scientific imperativese, essentially unrelated to further concerns and not essentially historical.

2.1 Interpretation of the First Passage

Although there are some differences between the ideas expressed in these three passages, I do not believe that Kuhn is inconsistent or that he later changed his position. I rather believe that these passages describe different aspects of the same or very similar doctrines, whose implications either he did not perceive clearly, or which, when he did perceive them, he tended to reject, in part no doubt because of the critical reactions by those who objected to what was rightly viewed as a strong challenge to the reigning realist approach to philosophy of science.

The first, extremely interesting passage suggests that under certain conditions, basic scientific advances change the way we view nature, or the mind-independent external world we seek to know, but also change nature itself. In denying that nature is fixed, or permanent, the passage rather clearly indicates that nature itself, or at least what we mean by 'nature,' is a historical variable, a mere 'construct,' which is not mind-independent, nor even the same for all qualified observers, although we may choose to treat it as such. This suggests that such knowledge as we have, including scientific knowledge, is arrived at by finite human beings who in the process of studying the world and themselves 'construct' both as a consequence of their very theories. The entire intellectual tradition is nothing more than the history of this process in the course of which knowledge of the world and the world – that is, the knower, the known and what is known about – are all constructed within the historical process itself.

There is an obvious distinction between studying mind-independent nature with the aim of determining its intrinsic characteristics, including the differences between our perceptions and what lies beyond them, and in some sense 'constructing,' 'producing' or 'creating' it. Kuhn's view of scientific progress does not appear to suggest that a change in a general, or non-scientific, perspective allows a scientific discovery to be made. It rather suggests that basic scientific advances, which may themselves depend on a change in scientific perspective, lead to a new conceptual framework, which in turn changes the way we view nature, which is itself a scientific 'construct,' not independent of, but rather squarely dependent on, theories about it. On this view, the familiar distinction between nature as it is perceived and nature as it is in independence of us cannot be made out. What we perceive as nature is at least partly a product of our scientific theories, which leads us to interpret what we encounter in experience in one way or another on that basis. To take an example which Kuhn himself discusses, combustion can be, and in fact is, interpreted in terms of oxygen if and only if oxygen is a known explanatory factor.

There is an elementary difference between the way the world is in independence from us and whatever we perceive. There is no way to know that we perceive the world as it is, and Kuhn does not claim that there is. One might, for instance, claim that our view of the unperceived independent natural world is itself a construct in either of two senses. One is that at different times and places different views are held by the wider community about the way the world is. Collingwood calls attention to the difference in the world views that dominate in different historical periods.[11] It seems obvious that the ancient Greek world which relied on fate and the actions of the gods and goddesses who lived on Olympus as explanatory factors is very different from the world in which we now live where explanation is based on causal analysis. Such differences further occur within science itself, for instance in the change from the pre-Copernican to the Copernican astronomical model, or again in the transition between the Newtonian and the relativistic models.

Scientific world views frequently influence philosophy, and less frequently the general view of ourselves and the world. Kant, who famously claims that all knowledge begins in, but is not limited to, experience, makes unverifiable assumptions about Aristotelian logic, Euclidean geometry, and about Newtonian mechanics as correctly representing reality as part of the critical philosophy. This attitude is no longer plausible after the discovery of non-Euclidean geometry, which destroys the presumed capacity to provide an exhaustive *a priori* analysis of the conditions of real experience.

It is reasonable to think that, like few or perhaps no other scientific discoveries, the introduction of Copernican astronomy forever changed the way we regard the world and ourselves, but also changed the world itself. A world in which the earth is no longer at the center of the universe belongs to a secular universe in which human beings no longer depend in any central way on God, in which revelation is no longer a primary source of knowledge, and in which the Church itself has no more than an occasional, but certainly not an essential, role to play.[12]

2.2 Interpretation of the Second Passage

The second passage suggests that, to use other terminology, which Kuhn also employs, a paradigm is either a world view (*Weltanschauung*) or a scientific world view (*wissenschaftliche Weltanschauung*). Kuhn's suggestion that a scientific paradigm is a world view links the former to the social context and to history. A world view is wider than a scientific world view since only some members of a given population are scientists. This difference suggests two further models of scientific activity. The first model, common now in the social constructivist approach to science, hints that no distinction, or at least no hard and fast distinction, can be drawn in practice between science and other forms of social activity. On this view, science is merely one kind of social activity.

In this respect, different attitudes are possible. Let us consider three of them. Paul Feyerabend is famous for his claim, which was perhaps not intended to be taken other than as a provocation, that voodoo is as good as quantum mechanics. This approach is arguably not intended to elevate voodoo, but rather to displace quantum mechanics from its supposedly permanent perch at the pinnacle of contemporary scientific knowledge. Another attitude is to deny, if scientific activity cannot be segregated from other forms of social activity, that science differs in kind from other forms of cognitive activity, or even that a separation can be drawn between scientists and the wider community. Thus Shapin and Schaffer argue that it is political sensitivity rather than epistemic success which led to Boyle's theory of the vacuum winnning over Hobbes's theory of the plenum in the seventeenth century.[13] On this model, science relies on extra-scientific factors which simply cannot be explained on purely intra-scientific criteria.

The other attitude toward scientific activity considers it as in principle and in fact isolated from the wider community. On this model, which is widely held, in fact deeply embedded within the scientific community, scientists formulate their theories, later elaborate them, and eventually discard them for reasons that are at most contingently related to the views of the members of the wider community, but directly related to the views of the members of the scientific community.

2.3 Interpretation of the Third Passage

The third passage is a version of an idea already stated in the Postscript to *The Structure of Scientific Revolutions* that does not encourage, but rather discourages a historical view of science and knowledge through a pluralist form of realism, close to Kant and to Putnam's internal realism, which stresses metaphysical realism in general, hence denies all forms of constructivism. Kuhn now presupposes, in order to avoid solipsism that would result if Davidson's objection to his supposed view of radical incommensurability were valid, that we share a common world. On this basis, he distinguishes in a loosely Kantian way between sensations, which we perceive, and stimuli, which can be said to cause them, but which are not themselves perceived. He says that we know with assurance three things: different stimuli can cause the same sensations, the same stimulus can produce different sensations, and sensation at least partly depends on education.[14]

As concerns the relation of stimuli to sensations, Kuhn appears to be making two related claims. On the one hand, sensation in general depends on the individual as well as on the individual's education. Different individuals perceive differently. One can further be taught to pick out different sensations, including those which are not clearly perceptible to an uneducated person, as an astronomical observer can be taught to distinguish six degrees of stellar magnitude with the naked eye. Since different individuals react differently to different stimuli, a univocal reading of the link between a given stimulus and a given sensation is not possible. With this in mind, Kuhn suggests that we can give different interpretations to one shared, mind-independent external world.

Note that the presupposed common world is situated on the level of stimuli, or perception, not on the level of sensations. In this passage, Kuhn is suggesting that we share one world as the common source of all experience, and about which we theorize in different ways, but which we do not know or at least do not know we know. In this respect, his view is close to Hilary Putnam's anti-relativist, pluralist idea of internal realism, since abandoned by its author, according to which different theories are possible about the same independent external world.[15]

3 Different Worlds Revisited

The work of this chapter has so far consisted in interpreting three different passages which concern the doctrine of different worlds. Each passage points in a different direction. The first points toward different worlds which exist before and after a fundamental scientific discovery. The second picks out the difference between different conceptions of paradigm as relating to the view of the wider community at a given time and place or, on the contrary, as confined to its scientific subset. The third indicates that there is a single mind-independent reality, or fixed nature, to which different observers react differently.

What is Kuhn's view of different worlds? It is not only charitable but, I believe, correct to credit Kuhn, not with three different views, but with a single theory which for whatever reason he was never able to formulate in satisfactory fashion and whose consequences he may not have clearly understood. These three passages from different periods in his work emphasize different aspects of a single complex, vaguely Kantian view, which can be reconstructed as follows. There is a common, mind-independent external world, which we can never know that we know as it is, but which we study in various ways in the cognitive disciplines. This same common world yields a single common set of stimuli, which are the same for everyone, but which are interpreted differently on the perceptual level by different observers. In reaction to fundamental scientific discoveries, observers shift to new and irreducibly different paradigms, which are evaluated on intra-scientific and extra-scientific grounds, but which refer to different worlds.

Interesting here is that, although Kuhn's statements on different worlds can be taken as suggesting that new theories literally 'construct' new worlds, hence modify reality which turns out not be invariant, as metaphysical realists believe, but a historical variable, he staunchly resists this implication in later clearly suppporting what looks like a pluralist form of realism. In that sense, Kuhn exemplifies the Kantian idea of an original thinker, who has an interesting new idea with respect to science, but is unable to work it out by himself.[16] Although Kuhn does not elaborate his idea, there is no reason why we should not regard it as an ongoing project, which can be taken beyond the use Kuhn makes of it in trying to understand its implications for a historical approach to knowledge.

Kuhn's view of different worlds raises a series of further questions about the nature and justification of scientific claims for cognitive objectivity, whose defense depends on the view of the cognitive object. If the world is a fixed nature, and we know it, as in the Platonic view of reality, then the problem of knowing that we know does not even arise. It arises, however, on a Kantian view, according to which the world is independent of any conceptual framework, but cannot be grasped directly. It becomes more pressing in a 'constructivist' perspective in which, since the world literally depends on conceptual schemes about it, a distinction cannot be drawn, or cannot be clearly drawn, between scheme and content.

Kuhn, who denies the very idea of a neutral standpoint,[17] rejects Popper's view of the falsification of scientific theories in favor of a roughly Peircean view of the assent of the relevant community.[18] Unlike Quine, for whom theories simply cannot be rejected on mere empirical grounds, since further adjustment is always possible, Kuhn favors an empirical constraint in that broadly based theories, or paradigms, can be 'refuted' by being confronted with problems for which they have no easy or even no acceptable response, or for which better responses are available.

How is the decision taken to accept or reject a particular paradigm, and as a result a particular world? In this respect, Kuhn, who is of two minds, regards science as continous, but also as discontinuous, with life. His ambiguity reflects, I believe, his inability to distinguish between a view of science as an interlocking collection of theories which do not discover but rather 'construct' the world, and a competing view of science as simply 'discovering' the world as it is, or at least as it is for us.

This difference is obviously related to the well-known distinction between the context of discovery and the context of justification, where the latter is further divided into personal and social elements on the one hand, and strictly scientific elements on the other. Someone who thinks that scientists 'construct' what they know is likely to acknowledge the relation of the theoretician to the surrounding context in a way that someone who thinks that scientists 'discover' the mind-independent world is unlikely to concede.

To begin with, in the initial formulation of his position, Kuhn favored the input of extra-scientific factors leading to a view of science as continuous with life, which is a main source of his reputation as a social epistemologist. It is characteristic of social epistemology that claims to know, like the people who make them, have an ineliminable social dimension. They are not, for instance, justified on either transcendental or empirical grounds, but rather in relation to the wider perspective, angle of vision, typical beliefs, in short the world view which prevails in a particular group at a particular time. An example might be the Copernican heliocentric view which, when it was introduced, did not explain the facts as then known any better than the rival Ptolemaic geocentric view.[19] This type of approach, which is favored by social constructivists,[20] such as the members of the Edinburgh group,[21] is typically rejected by non-constructivists or anti-constructivists as substituting relativism, skepticism, and naturalism for philosophical science, as in Husserl,[22] and by philosophers of science, for instance Canguilhem, as replacing philosophical critique through social psychology.[23] In the view introduced in the Postscript, where he stresses the more conventional idea of science as 'discovering' the mind-independent world Kuhn suggests on, the contrary, that scientific claims to know are justified on intra-scientific grounds much closer, say, to Quine's epistemic holism.

What is the difference? Both forms of justification are broadly pragmatic, but in different ways. A sociological view of justification as cohering with a wider world view, as in William James's theory of truth, suggests that scientific theories cannot be self-contained, that is, worked out solely within the scientific framework, since they must always satisfy extra-scientific criteria in the course of their adoption. This reading of Kuhn's view is supported by his adoption in the Postscript of the term 'disciplinary matrix' instead of 'paradigm' to refer, *inter alia*, to a shared commitment to basic beliefs and common values, which is typical of a *Weltanschauung*. Such criteria point toward extra-scientific factors ingredient in the adoption of a basic alteration in the prevailing scientific framework. The rather different reading of a paradigm as an exemplar, or shared example, suggests that a new view provides a different way of resolving outstanding scientific problems, whose success can be evaluated on purely intra-scientific grounds without any further claims about the status of the world as, to put the point provocatively, either 'found' or 'made.'

It obviously makes a difference if a paradigm, or so-called world (as in different worlds), is regarded as continuous with society as a whole, hence subject to evaluation through extra-scientific factors, or as discontinous with society, hence solely and wholly explicable on intra-scientific grounds. Carnap, an early proponent of the reductionist view that theories are, or at least in principle can be, seamlessly derived from facts,[24] was committed to the idea that there is a mind-independent external world and that it can be known. He was famously countered by Neurath's claim that there are no ascertainable pure, non-metaphysical facts,[25] which is compatible with the idea that there is a mind-independent external world and with the very different idea that the world is theory-dependent. Neurath's claim is continued in Quine's anti-reductionist view that theories can never be directly derived from facts.[26] If Quine is correct, then the so-called explanatory gap can be filled in at least two ways: on sociological grounds through the coherence of the new view with views already held, or through the capacity of the new view to solve outstanding problems. The former is neutral with respect to the status of the world, although the latter at least implies that the cognitive object is simply independent of the knower. At different times Kuhn suggests both conceptions, earlier the former, and later the latter.

In this respect there are three possibilities. One is to suppose that Kuhn later retreated from his earlier, more sociological theory. To a certain extent that seems to be the case. His later writings tend to stress intra-scientific factors, hence a purely scientific understanding of scientific theory, more than sociological factors. In this respect, it is fair to say that he later changed his position. A second possibility is that Kuhn is simply inconsistent since one cannot simultaneously adhere to sociological, or to extra-scientific, and to intra-scientific factors with respect to paradigms. Those who think there is a clean choice in science, which is either one or the other – that is, either only sociological or only non-sociological – will be impressed by this alternative. There is finally a third possibility, that is, that both sociological and a-sociological, strictly scientific factors can and often do play a role in the adoption of a wide-ranging new theory, hence in the transition from one world to another. To return to the Copernican example, the sociological factors, which were prominent early in the transition away from Ptolemaic and toward Copernican astronomy, were later supplemented by more clearly intra-scientific factors due to Kepler and then especially to Galileo.

Unfortunately, Kuhn does not bring the sociological and the scientific factors together in a larger synthesis. His failure to do so notoriously led some of his readers to believe that he slighted scientific objectivity in favor of extra-scientific factors.[27] On the contrary, I believe that much of the interest of Kuhn's view lies in its acknowledgment that science can neither be reduced to the general surrounding social and intellectual context, nor separated from it, precisely the approach one would expect from someone unable to choose between a view of science as 'constructing' and as 'discovering' the world. The effort to do either arguably misdescribes what in fact occurs, hence misdescribes the scientific process. Kuhn stands out in his attention to both factors. Yet in his wake there is still no satisfactory effort to combine them both in a large theory.

3.1 Different Worlds and Scientific Progress

A further difficulty, on a 'constructist' hence anti-realist view of our grasp of the world, lies in grasping scientific progress which cannot simply be understood on some version of the familiar correspondence view. According to Kuhn, the progression from the old paradigm to the new one makes propress in solving recognized puzzles[28] and in refuting past paradigms,[29] but cannot be known to progress toward anything, such as in knowing the way the mind-independent external world is.[30] This is the view one could reasonably expect from him if he favored the idea that the world is no more than a 'construct,' but also if, like Cassirer, he held that the relation of scientific theory to the mind-independent external world is at best symbolic, but not, say, representational.[31] Both views suggest that our relation to world cannot simply be one of 'discovering' it as it is.

This raises the prospect that the history of science is no more than the complicated tale of wandering over the centuries among the many different worlds 'constructed' by different observers over the centuries, a series of worlds none of which is any better than the rest, since science itself is not cumulative, in the way it would be if it were a way of discovering the way the world is as it is. The difficulty, then, if one favors a 'constructivist' view, is to make sense of the idea that science progresses through theory change, since in general later, more recent theories are better than their earlier predecessors.

The dismal prospect that the price of constructivism and even of anti-realism is scientific stasis, the natural scientific equivalent of *philosophia perennis*, has been identified and criticized in a recent paper by Steven Weinberg, the well-known physicist. According to Weinberg, unless we are getting closer and closer to a grasp of the way things are in independence of us, in short, unless we are engaged in an activity which in fact 'discovers' mind-independent reality, science will turn out to be an irrational activity. [32] In other words, put very simply the objection runs: either science is cumulative, or it is irrational. Weinberg has clarified his criticism in a recent letter in writing: 'If I agreed with Kuhn's judgment about the progress of science, that there is no sense in which science offers a cumulative approach to some sort of truth, then the whole enterprise would seem rather irrational to *me*, even if not to Kuhn.'[33]

In part because of his rejection of metaphysical realism, objections of relativism and irrationalism are often raised against Kuhn, for instance by Popper[34] and Lakatos.[35] Weinberg is not accusing Kuhn of favoring irrationalism, but rather of proposing an irrational view of science. Like Weinberg, Kuhn regards science as the best modern example of rationality.[36] Weinberg's objection raises two problems, which can be stated as questions: Is science progressive, and if so, in what way? Does science reach truth?

For Weinberg, science is progressive in that it tends toward a final theory, which will be constructed at some future time on the basis of what is previously known. Such a theory would bring together optics, magnetism and electricity as in Maxwell's time, as well as electrodynamics and other theories of elementary forces which currently form the Standard Model of elementary particles, in a single unified theory.[37]

Weinberg's idea of science is obviously too simplistic for two reasons. First, there is simply no way to know that we are getting close or even closer than before to representing the way the mind-independent world is, since, as Kant already pointed out, there is obviously no way to know how it is in itself. There is no known way to cash out assertions to grasp independent reality. Yet this problem is obviously neutral about the relation of theory to the world, which can be either 'constructed' or 'discovered.' Second, there is no way to know that as current puzzles are resolved others will not appear in their place, which is also neutral to the same relation. Despite the impressive accomplishments of scientists, we simply cannot know that, as some think, we are down to the details now.[38] For there is not currently, and apparently never will be, any way to estimate how close science is to its final state.[39]

Can we know that we are getting closer to the truth? Although Peirce is sometimes read in this way, I believe it is incorrect to hold that science is demonstrably getting closer to a final account of the world, however understood. The problem is already clear in Kant. In a famous letter to Herz, Kant, who was an empirical realist, suggests that an adequate theory of knowledge must explain the relation of the representation to the (independent) object.[40] But we cannot do so since there is simply no way to know that we are getting close or even closer than before to representing the way the mind-independent world is.

Is science true? Is it cumulative? If 'true' means something like knowing the way the world is in itself, say through a correct representation of mind-independent reality, then science cannot reach truth. Yet science is obviously cumulative, hence progressive. Apparently following Lakatos' view of a research program as featuring a series of theories, and not one given theory within the methodology of research programs,[41] Weinberg suggests an idea of science as composed of hard and soft parts. The former refers to that part of scientific theories which does not change, but remains as a permanent accomplishment, such as equations and so on, and the latter refers to what does change, for instance the belief in ether, and so on. Yet this cannot be correct. Other than an interest in nature, Newtonian mechanics, which is quantitative, has nothing obviously in common with Aristotelian physics, which is qualitative. And relativity theory does not take over parts of, but rather substitutes a different, conceptual framework for, Newtonian mechanics.

4 Conclusion: Different Worlds and Historical Progress

Paradoxically, since Kuhn is a fine historian of science, his philosophy of science lacks an explicitly historical dimension toward which it points. In a word, he offers a history of science that is paradoxically non-historical. Yet his view of revolutionary changes in science points toward a historical view of knowledge, including science, which he captures in his deep suggestion that at different historical moments we live in different worlds. Like the later

Wittgenstein and other contextualists, Kuhn understands that claims to know are justified in relation to conceptual frameworks.[42] Unlike Wittgenstein, who is a basically ahistorical thinker, Kuhn clearly sees that conceptual frameworks succeed each other as a function of their relative capacity to solve outstanding scientific puzzles relative to different worlds which they literally call into being. Different conceptual frameworks are literally, as Kuhn suggests, different worlds which, for a time, as long as they continue to hold conceptual sway, scientific observers contingently inhabit, and which determine the world. It follows that the world one contingently happens to inhabit, and on the basis of which one observes and reflects, is inseparable from a given historical moment, hence dependent on human history. In a word, different worlds are historically relative.

It is sometimes urged that either there is the realist's one world and we know it, or there are the relativist's many worlds which cannot be known. This seems doubly false. On the one hand, it is plausible that, as van Fraassen, suggests, our scientific views are formulated or chosen to fit the facts as we know them,[43] but that we cannot know how they relate to the way the world is. To be sure, this is only part of the story if, as I believe, so-called facts are dependent on conceptual frameworks. Van Fraassen's idea is compatible with the view that through our theories we 'construct' the world. Yet it is not possible to show, as Putnam earlier claimed, that there is one single world which we in fact know from different points of view. On the other hand, since the very idea of the independent external world is itself a historical variable, its existence, at least as Putnam apparently intends it, seems doubtful. Different stages in the history of science simply cannot be shown to offer different interpretations of the same independent external world; they rather propose different views of different independent external worlds, or the world as it is understood in different historical moments. Not only do our views of the cognitive object change as our theories change, but even the cognitive object, which depends on the theories we formulate also changes.[44] I conclude that we can safely deny the disjunction; for we cannot know that we know the single independent world, which is a mere posit, or even that it exists, which is another posit; but we do know the relativist's many worlds, which alone are compatible with scientific progress.

Notes

1 See Putnam (1981), p. 49.
2 See Quine (1960), pp. 22–3.

3 Davidson's critique of the idea of incommensurability leads him to the absurd consequence of rejecting any and all conceptual frameworks. See Davidson (2001).
4 See Kuhn (1970), Ch. 12.
5 Ibid., p. 148.
6 Ibid., p. 150.
7 Ibid.
8 Ibid., p. 118.
9 Ibid., p. 175.
10 Kuhn (1977), p. 309, n. 18.
11 See Collingwood (1965). This claim is disputed by Lakatos (1978), Vol. I, p. 174.
12 For discussion, see Dewey (1960).
13 Shapin and Schaffer (1985), p. 342.
14 Kuhn (1970), p. 193.
15 Putnam (1981).
16 See Kant (1961), p. 654.
17 See Kuhn (1970), p. 126.
18 See ibid., pp. 94, 216.
19 See Kuhn (1957).
20 This term, which is used in many different ways, is associated with authors such as Scheler, Mannheim, Berger and Luckmann, Merton and others.
21 See, for example, Barnes, Bloor and Henry (1996); see further Collins (1992).
22 See Husserl (1965).
23 See Canguilhem (1994), p. 46: 'For [Kuhn] a paradigm is the result of a choice by its users. Normal science is defined by the practice in a given period of a group of specialists in a university research settting. Instead of concepts of philosophical critique, we are dealing with mere social psychology. This accounts for the embarassment evident in the appendix to the second edition of the Structure of Scientific Revolutions when it comes to answering the question of how the truth of a theory is to be understood.'
24 Carnap (1967).
25 See Neurath (1959).
26 See 'Two Dogmas of Empiricism,' in Quine (1961), pp. 20–46.
27 See Laudan (1990).
28 See Kuhn (1970), p. 166.
29 See ibid., p. 167.
30 See ibid., pp. 170–71.
31 See Cassirer (1923–29), Vol. 3, p. 20. According to Cassirer, Herz was the first scientist to make the decisive turn from a copy-theory to a symbolic view of science.
32 See Weinberg (1998).
33 See Weinberg's letter in the *New York Review of Books* (Weinberg, 1999).
34 See Popper (1970), p. 56.
35 See Lakatos (1970), p. 93, where he claims that 'For Kuhn scientific change … is a mystical conversion which is not and cannot be governed by rules of reason and which falls totally within the realm of the *(social) psychology of discovery*.'
36 Kuhn, who is sometimes linked with irrationalism, in fact rejected irrationalism in his view of science as the best current example of rationality. In 1970, he wrote: 'I do not for a moment believe that science is an intrinsically irrational enterprise … I take that assertion not as a matter of fact, but rather of principle. Scientific behavior, taken as a whole, is the best example we have of rationality' (Kuhn, 1971, pp. 143–4).
37 See, for example, Weinberg (1993).

38 See Horgan (1996).
39 See Maddox (1998).
40 See letter of 21 July 1772, in Kant (1967), pp. 70–76.
41 See Lakatos (1970).
42 See Wittgenstein (1972).
43 For the view of constructive empiricism, see van Fraassen (1980).
44 For an argument of this kind, see the 'Introduction' to Hegel (1977), pp. 46–57. For discussion, see Rockmore (1998), pp. 23–36.

References

Barnes, Barry, Bloor, David and Henry, John (1996), *Scientific Knowledge: A Sociological Analysis*, Chicago: University of Chicago Press.

Canguilhem, Georges (1994), *A Vital Rationalist: Selected Writings from Georges Canguilhem*, trans. A. Goldhammer, ed. F. Delaporte, New York: Zone Books.

Carnap, Rudolf (1967), *The Logical Structure of the World*, trans. Rolf A. George, Berkeley, CA: University of California Press.

Cassirer, Ernst (1923–29), *The Phenomenology of Knowledge, The Philosophy of Symbolic Forms*, Vol. 3, trans. R. Manheim, New Haven, CT: Yale University Press.

Collingwood, R.G. (1965), *The Idea of Nature*, New York: Oxford University Press.

Collins, Harry (1992), *Changing Order: Replication and Induction in Scientific Practice*, London: Sage.

Davidson, Donald (2001), *Inquiries into Truth and Interpretation*, 2nd edn, Oxford: Oxford University Press.

Dewey, John (1960), *The Quest for Certainty*, New York: Putnam.

Hegel, G.W.F. (1977), *Phenomenology of Spirit*, trans. A.V. Miller, New York: Oxford University Press.

Horgan, John (1996), *The End of Science*, Reading, MA: Helix Books.

Husserl, Edmund (1965 [1910]), 'Philosophy as Rigorous Science', trans. in Q. Lauer (ed.), *Phenomenology and the Crisis of Philosophy*, New York: Harper.

Kant, Immanuel (1961), *Critique of Pure Reason*, trans. N.K. Smith, B 862, London: Macmillan.

Kant, Immanuel (1967), *Philosophical Letters, 1759–99*, trans. Arnulf Zweig, Chicago: University of Chicago Press.

Kuhn, Thomas S. (1957), *The Copernican Revolution: Planetary Astronomy in the Development of Western Thought*, Cambridge, MA: Harvard University Press.

Kuhn, Thomas S. (1970), *The Structure of Scientific Revolutions*, Chicago: University of Chicago Press.

Kuhn, Thomas S. (1971), 'Notes on Lakatos', in Roger C. Buck and Robert S. Cohen (eds), *Proceedings of the 1970 Biennial Meeting, Philosophy of Science Association, Boston Studies in the Philosophy of Science*, Vol. 8, Boston: D. Reidel, pp. 143–4.

Kuhn, Thomas S. (1977), *The Essential Tension: Selected Studies in Scientific Tradition and Change*, Chicago: University of Chicago Press.

Lakatos, Imré (1970), 'Falsification and the Methodology of Scientific Research Programs', in I. Lakatos and A. Musgrave (eds), *Criticism and the Growth of Knowledge*, Cambridge: Cambridge University Press, pp. 91–195.

Lakatos, Imré (1978), *Philosophical Papers*, 2 vols, ed. L. Worrall and G. Currie, Cambridge: Cambridge University Press.

Laudan, Larry (1990), *Science and Relativism: Some Key Controversies in the Philosophy of Science*, Chicago: University of Chicago Press.

Maddox, John (1998), *What Remains to be Discovered?*, New York: Free Press.

Neurath, Otto (1959), 'Protocol Sentences', in A.J. Ayer (ed.), *Logical Positivism*, New York: Free Press, pp. 199–208.

Popper, Karl (1970), 'Normal Science and its Dangers', in Imré Lakatos and A.E. Musgrave (eds), *Criticism and the Growth of Knowledge*, Cambridge: Cambridge University Press.

Putnam, Hilary (1981), *Reason, Truth and History*, New York: Cambridge University Press.

Quine, W.V.O. (1960), *Word and Object*, Cambridge, MA: MIT Press.

Quine, W.V.O. (1961), *From A Logical Point of View*, New York: Harper and Row.

Rockmore, Tom (1998), *An Introduction to Hegel's Phenomenology of Spirit*, Berkeley, CA: University of California Press.

Shapin, Steven and Schaffer, Simon (1985), *Leviathan and the Air-pump: Hobbes, Boyle and the Experimental Life*, Princeton, NJ: Princeton University Press.

van Fraassen, Bas C. (1980), *The Scientific Image*, Oxford: Clarendon Press.

Weinberg, Steven (1993), *Dreams of a Final Theory: The Search for the Fundamental Laws of Nature*, New York: Vintage.

Weinberg, Steven (1998), 'The Revolution that didn't Happen', *The New York Review of Books*, 45 (15), 8 October, pp. 48–52.

Weinberg, Steven (1999), letter in the *New York Review of Books*, 46 (3), 18 February, p. 49.

Wittgenstein, Ludwig (1972), *On Certainty*, ed. G.E.M. Anscombe and G.H. von Wright, New York: Harper.

Chapter 8

Incommensurability Modestly Recovered

Joseph Margolis

1

The most important fact about incommensurability is how blandly it is ignored and casually denied. If you scan the entire range of current English-language philosophy, hard-core analytic philosophy or something less decisive, you will find that incommensurability is conceptually well nigh invisible. At the risk of a bad joke, I would say that it's a little like racial prejudice viewed through the eye of a satisfied liberal: it does not exist because it would be incompatible with our legal assurance of equal opportunity – 'which no one can deny.' But that means, of course, that incommensurability is essentially a political or ideological problem for the dominant currents of recent American philosophy. Or, better, it needs to be analyzed as if it were (if its proper challenge is to be grasped at all).

The question is a conceptual one all right, but the trick is to help those who can't even see what must be seen to get their proper bearings. For you realize that the familiar, much-trafficked philosophical terrain still harbors mysterious practices – right under our nose. The best philosophies are often occupied with making the obvious obvious.

The latter-day saints of incommensurability – or, better, incommensurabilism – are, of course, Thomas Kuhn and Paul Feyerabend. Both were intuitively gifted, but neither was politically adept in the required academic way. Kuhn was deficient in dialectical skills and was even temperamentally opposed to his own argument; and Feyerabend was, finally, too antic to care about making his argument effective. The result was that both were outflanked, though they made their case naggingly enough. It's been officially written off, you realize, most famously by Donald Davidson, so cunningly in fact that no one now admits the blind sight of our polite denials. There's the charge at least.

It is still worth recovering, except that if the charge is valid then much of recent canonical philosophy may have victimized itself, may have failed to perceive the profound distortion of its sincerest practice. Seen that way, our usual inquiries cannot be trusted to mean what they are said to mean: we shall need all sorts of indirection to see what we are asked to see. I warn you, therefore, that we cannot expect to rely on the corrective power of instant intuition, though it remains true enough that we must finally fall back to something like intuition. In the last analysis, things either fall into place happily or they do not – or so we say!

Start, then, with an oblique clue. Hilary Putnam, who yields to no one in his contempt for relativism and incommensurabilism, asserts unblinkingly: 'as soon as one tries to state relativism as a *position* it collapses into inconsistency or into solipsism (or perhaps solipsism with a 'we' instead of an 'I'). The thought that everything we believe is, at best, only 'true in our language game' isn't even a coherent thought: is the very existence of our language then only 'true in our language game'?[1]

That is a very slippery line of reasoning. Putnam sides with the argument pointedly brought in Plato's *Theaetetus* against Protagoras; for Putnam holds that relativists and incommensurabilists are either solipsists 'with a "we"' or solipsists 'with an "I"' – where, that is, the latter formula is assigned to Protagoras, though the former might be Putnam's preferred description of Protagoras as well.

I grant the force of the imputed charge, though not the charge itself: I agree entirely with Putnam – and with Plato, *if* Plato did believe that what he dramatized in the dialogue was the best reading of Protagoras' position (which I do not accept). I agree that, thus construed, relativism and incommensurabilism would be hopelessly inconsistent and incoherent. Fine. But you cannot fail to remark that the divergence, or incompatibility, or even the incommensurability, of the beliefs and convictions of different individuals and different societies *does not entail* that those who differ thus subscribe to the 'relationalist' formula (Socrates', in the *Theaetetus)*, construed either as a solipsism 'with an "I"' or 'a "we",' the formula, that is, that holds that 'true means "true-for-*x*"' (where '*x*' is either an individual or a society).[2] Putnam has gained his point in the sparest way: he admits no obligation to examine other possibilities.

'Solipsism with a "we"' is not even a true solipsism, of course, though it is a fatal mistake for societies that interact with one another (and are aware that they do). Bear in mind that there is no known language for which there are no apt bilinguals. Nevertheless, the mistake may be no more than a naive conviction. It rises to the level of deep incoherence only if it is a deliberate philosophical presumption on which an entire body of would-be truths is thought to depend. We are not driven to any such premise. It was indeed Evans-Pritchard's presumption regarding the Azande, which, rather memorably, Peter Winch undertook to examine.[3] It was also the presumption of Ranke's very different, but comparably incoherent, historicism.[4] Curiously, Putnam overlooks the possibility that judgments and truth-claims may diverge, may be incompatible, may even be incommensurable, without the opposed parties ever being committed to divergent, incompatible, or incommensurable accounts of truth, *or* ever committed to the relationalist theory of truth in any form.[5]

It is not in the least clear that Putnam ever admits the possiblity. It would of course go contrary to the spirit of his *Grenzbegriff*, mentioned only once at the very close of his best-known book, *Reason, Truth and History*, which suggests that he does not allow for any such possibility.[6] He offers only this unguarded (and unexplained) dictum – in the context of airing relativism's prospects: 'The very fact that we speak of our different conceptions as different conceptions of *rationality* posits a *Grenzbegriff*, a limit-concept of the ideal truth.'[7]

But there is no known compelling argument to the effect that subscribing to 'different conceptions of rationality' (or truth) *entails* the incomparability or unintelligibility of such conceptions. Putnam *never* recognizes the problem, which bids fair to undermine his entire argument. It also undermines, as we shall see, Donald Davidson's extraordinarily influential claim. For Davidson is the acknowledged architect of the widespread judgment that incommensurabilism is and must be utterly incoherent – beyond redemption. And yet, of course, there is nothing intrinsically incoherent about a contest between incompatible or incommensurable doctrines. If that is so, then incommensurabilism may prove viable after all; and if it is, then its coherence will expose a very deep – an almost invisible – mistake in our present dismissal of Kuhn and Feyerabend.

Please notice that, in putting the matter this way, the fate of incommensurabilism (and relativism) cannot be made to depend on Putnam's flawed analysis of relativism or Davidson's flawed analysis of incommensurabilism, or (for that matter) on Kuhn's or Feyerabend's flawed analyses of particular episodes in the history of science. All that is needed are reasonable specimens or proposals of what to count as relativist or incommensurabilist options. My aim here is to draw attention to neglected conceptual possibilities that cannot rightly be ignored if the usual realist options are admitted. The reason, obviously, is that we cannot segregate the troubling options by any satisfactory criterial rule so that the threat of relativism and incommensurabilism (or historicism and constructivism) can be kept from sight until the central cases are disposed of in the approved canonical way. No. The point is that the threat of the deviant options bleeds into the fate of the central cases themselves.

Putnam's formula, a 'solipsism-with-a "we",' is, effectively, a clever summary of Davidson's more ramified attack on Kuhn's and Feyerabend's incommensurabilism – the nerve of Davidson's well-known essay, 'On the Very Idea of a Conceptual Scheme.'[8] Putnam offers the same charge in a Wittgensteinian idiom that he invents for the purpose, which, once again, betrays the inexplicable blind spot already remarked. Glossing Wittgenstein, he says very neatly: 'To say something is true in a language game is to stand outside of the language game and make a comment: that is not what it is to play a language game.'[9] Of course, he's right – *on* the reading he advances. But it's hardly true that '*to* say something is true in a language game *is* to stand outside of that language game and make a comment' – in the sense of standing outside all the resources of the language in which this or that 'game' is played! To say that this or that is true, say, speaking as the Masons speak – *is not* to say that 'true' *means* 'true in Masonic talk' (even if what is said *is* true according to Masonic jargon). No. There's a blind spot there.

The key is this. Putnam claims to be a pluralist, by which he means to say – denying any single, uniquely valid 'God's Eye point of view' – that 'realism is *not* incompatible with conceptual relativity.' He insists, in effect, that this sort of pluralism does not entail anything like the Protagorean formula of the *Theaetetus*[10] or the objectivist view that we *do* know the real world as it is apart from our inquiries and know that there is only one finally valid account of its properties. He intends his pluralism as a third way. He's touched on something quite important here. But it cannot help him in his own argument.

There are two disastrous weaknesses in what Putnam says. For one, if he's right about his pluralism, then, contrary to what he says, the same will be found to hold for incommensurabilism: realism *is* compatible with incommensurabilism if, as may be shown, it requires no more than that what Putnam means by 'conceptual relativity' extends to alternative epistemologies as well. That is the stunning challenge Kuhn and Feyerabend put their finger on. It's been completely overlooked by Putnam and Davidson, which means in effect that it's been completely overlooked in the analytic literature. The other weakness is just as telling: Putnam (also Davidson) never broaches the epistemological issue on which, alone, his own 'third' option depends and (as he supposes) may be shown to be relevantly different – that is, different in a justificatory way – from the incommensurabilist alternative. The connecting premise is simply missing.

The evidence is buried in the following thought-experiment:

> If the sentence, 'points are mere limits' is a contrary of the sentence 'points are not limits but parts of space, [where each] sentence occurs in a systematic scheme for describing physical reality ... *though the two schemes are in practice thoroughly equivalent*, then [it may seem as if] we are in trouble

> … But the whole point of saying that the two schemes are in practice thoroughly equivalent is that, far from leading us to incompatible predictions or incompatible actions, it makes no difference to our predictions or actions. [Furthermore, a given] sentence in one [scheme] can be correlated in an effective way with a 'translation' in the other scheme, and the sentence and its translation will be used to describe the same states of affairs.[11]

There is nothing wrong with Putnam's proposal – except this: How does Putnam know that the two sentences describe 'the same states of affairs'? Can he really assure us that there are no epistemologically incommensurable schemes – *never* are – so that we can always in principle decide whether we are or are not addressing 'the same state of affairs'?

That is the precise point of Kuhn's and Feyerabend's challenge, the point of the puzzling Priestley/Lavoisier controversy. Kuhn and Feyerabend hold that there *are* such cases – which then require resolution by measures that cannot be canonical.

Putnam never reaches the question at all. For what he offers (which is interesting in its own right) is simply that, *for any accepted description of a given state of affairs*, we are always free to introduce *another* scheme of individuating or parsing the same world – another ontology, if you please – that can be made 'translationally' equivalent (Putnam's term, because it is Quine's term in the context of the *ur*-problem of *Word and Object*) to the sentences of the scheme with which we begin. Putnam never invokes the principle of the inseparability of metaphysics and epistemology, which, at the time of writing, was the leading theme of his own ill-fated 'internal realism.'[12] It is also, as it happens, a strategic theme favored in that important interval of philosophical discovery that spans Kant and Hegel and comes to rest in Hegel, in superseding the transcendental assurances on which Kant had relied in the first *Critique*. In short, Putnam ignores a theme that implicitly links his increasingly pragmatist version of realism with its unacknowledged Hegelian sources – which, in much the same way, informs Kuhn's and Feyerabend's view. Marvelous irony all around!

The post-Kantian discovery is far more important than Putnam's rejection of his own internal realism.[13] For, of course, it is the constructivist reading of the inseparability of metaphysics and epistemology, precluding as it does any assured or privileged fixity of 'the same states of affairs,' that ultimately provides the dialectical bite of Kuhn's and Feyerabend's incommensurabilism. Kuhn may not have been aware of the conceptual connection, and Feyerabend may not have been interested in acknowledging any such reliance; and both may be challenged in their reading of particular episodes in the history of science. But it is there all the same, and makes possible *their* thinned-down intuitions.

Although it was remarkably unguarded, Davidson's charge of incoherence[14] leveled against Kuhn's and Feyerabend's original challenge does begin to define the outer limits of what we should understand by incommensurability. Davidson neglects the common or garden varieties of incommensurability, those in fact that do not depend on the vexed epistemological doctrines so many have stumbled over; the common varieties actually help us to see what may be salvaged from the latter. Incommensurability remains, in general, a muddle, even though, setting Kuhn and Feyerabend aside, it cannot fail to strike us as, or as harboring, a perfectly respectable, even ubiquitous, even quietly acknowledged, phenomenon. The reason the entire issue was featured in the derogatory way it was in the analytic tradition is that it was mistakenly conflated with historicism's deeper challenge to the objectivist pretensions of the analytic canon: in effect, what, in the name of realism and naturalism, is hardly more than a late continuation of the fatal 'Cartesianism' that spans Descartes and Kant. Clearing the air on incommensurability should help us put the larger issue in a better light.

This needs to be explained. By 'Cartesianism' or 'Cartesian realism' I mean any realism, no matter how defended or qualified, that holds that the world has a determinate structure apart from all constraints of human inquiry and that our cognizing faculties are nevertheless able to discern those independent structures reliably. 'Cartesianism' serves as a term of art here, not confined to Descartes' doctrine. It ranges over pre-Kantian philosophy, Kant's own philosophy (quixotically), and over the views of such contemporary theorists as Putnam and Davidson.

Hegel is usually credited with formulating (in his *Phenomenology*) the most compelling sketch of a *reductio* of Cartesianism up to the early nineteenth century. The main feature of Hegel's strategy, which, in American philosophy, is preserved (almost without attribution) among the classic pragmatists (particularly Dewey), retires altogether the very idea of reference to a 'noumenal' world or a world the properties of which are separable from whatever *they* are said to appear to be to human inquirers, and reinterprets 'appearances' (*Erscheinungen*) as open to the recovery of no more than a 'constructed' realism, that is, without any recuperative use of the former 'Cartesian' habit of opposing or disjoining 'appearance' and 'reality' completely.

The opponents of 'realism' usually intend as their target some version of the 'Cartesian' doctrine. One must take care, therefore, in recovering the term, in agreement with the so-called 'anti-realists' (of various stripes), to appreciate both that some among them (Michael Dummett, for instance) are also 'realists' in a recuperated sense.[15] They are *not* 'idealists' in either the sense akin to Kant's (whether transcendentalist or not) or in the sense favored by certain German and British Idealists (Schelling or T.H. Green, for instance). They may remain Cartesians of a sort, as does Dummett, no less than Davidson. Otherwise, the recuperated sense, in accord at least with the central resolution of both the Cartesian and Kantian paradoxes offered in Hegel's *Phenomenology* (the original site of the full resolution of both paradoxes) must be constructive (or constructivist) which, for some commentators (Michael Devitt, for one[16]), is, erroneously, sufficient to convert the anti-realist into an idealist. 'Constructivism' signifies: (i) that there can be no principled disjunction between epistemology and metaphysics, or between the evidentiary roles of cognizing subjects and cognized objects in assessing the objectivity of truth-claims; and (ii) that whatever (constructively) is construed as ontically independent of human inquiry is, as such, epistemically dependent, in accord with (i). The great shift in modern philosophy ushered in by Hegel replaces transcendental constructivism (Kant), effectively an idealism, with a constructivism that depends on human resources and escapes idealism (Hegel). But if we abandon (as we must) all forms of cognitive privilege, admit the coherence of incommensurability, *and* historicize cognitive resources themselves, we cannot fail to see that we are but a step away from admitting the viability of Kuhn's and Feyerabend's incommensurabilism. Analytic philosophies – notoriously Davidson and Putnam – ignore these dialectical linkages.

The connection with the views of Kuhn and Feyerabend is simply that they, as opposed to Putnam and Davidson, subscribe to a realism (if we may call it that) that implicitly eschews Cartesianism, favors some form of constructivism, and does so on the basis of a historicized account of our cognizing abilities. In this regard, they are 'Hegelian' in spirit (though not explicitly), very probably more radical then Hegel, certainly more radical than the pragmatists, though in a way that extends the pragmatists' own critique of Cartesianism. Once you have these themes in hand, it is relatively easy to see how Putnam simply ignores the incommensurabilist option (assuming, that is, that incommensurabilism can be formulated coherently).

The rejection (as by Davidson) of incommensurability on grounds of its patent incoherence or paradoxicality or self-contradiction presupposes something akin to the thesis of a closed universe – most typically, a closed physical universe, as among the supervenientists, of whom Davidson and Jaegwon Kim are perhaps the best-known exemplars[17] – conceptually constrained at the very least by prior necessities of thought that simply rule incommensurability out of bounds. The supervenientist presumption – or, better, the closed-universe presumption on which supervenientism itself depends – shows very clearly that if we allow such conceptual limitations to constrain our account of reality *a priori* (or to count as metaphysical discoveries as such), we will be forced to argue in the Cartesian manner.

Davidson presumes that there *is* an obvious necessity *de cogitatione* that systematically precludes the bare conceptual eligibility of the incommensurability thesis. Allow that to stand unchallenged for the moment, though it is also completely undefended: it turns out to be the essential nerve of Davidson's objection. All modal detractors (as we may call them), those who agree with Davidson, argue that the defeat of incommensurabilism need never concern itself with anything like the historicity of thought and cognition or the historicity of a realism fitted to the sciences, which, of course, are featured in Kuhn's 'different worlds' conjecture.

What I reported a moment ago about Putnam's pluralism and his attack on 'solipsism with a "we"' could never have been directed against incommensurabilism (or relativism) if he had not, at some subterranean level of thought, supposed that there *was* a condition of conceptual closure that ruled out such historicized developments as those Kuhn and Feyerabend claimed to find. Make no mistake about it: Kuhn's and Feyerabend's account of incommensurabilism is an extreme epistemological option that cannot be matched by anything that may be found in Hegel or the pragmatists – *a fortiori*, by anything that may be found in analytic philosophy.

I see no other plausible explanation of Putnam's blind spot than his residual Cartesianism. It explains his appeal to his *Grenzbegriff*, and it is a corollary of Davidson's explicit objection against Kuhn.

Here, I must say, candidly: Davidson's presumption has never been successfully demonstrated, and rests on an utter misreading of Kuhn's original texts; also, against both Putnam and Davidson, the prospects of Kuhn's conjecture about paradigm shifts cannot be freed from the empirical evidence about epistemological incommensurabil*ism* in the history of science itself (as in Kuhn's famous example of the Priestley/Lavoisier controversy), and that, apart from incommensurabilism, the incommensurabil*ity* debate (the debate about ordinary conceptual incommensurability) should provide specimens that do not rely, philosophically, on the other. To admit all this, which seems eminently reasonable, yields an instant advantage. For, if conceptual incommensurabilities are uncontroversial and widespread, then it is but a step to make the case that epistemological incommensurabilism is likely to appear *within the range of ordinary incommensurabilities*! If so, then

Putnam and Davidson must be whistling in the wind. There's the argument that's needed.

You must remember that Kuhn took Galileo to be his principal specimen of a practicing scientist, someone actually familiar (as a result of his own training) with the very different 'worlds' of Aristotelian science and the emerging dynamics of late medieval science that led directly to Galileo's contribution regarding the pendulum. That is, Kuhn could never have consistently supposed that the fair admission of incommensurability entailed the unintelligibility of one such paradigm scanned by the partisans of another, or the incomparability of two such paradigms, or even the impossibility of translating the terms of either paradigm by way of an idiom that made room for the other. If you admit this much, you cannot fail to admit that a considerable part of the incommensurabilism dispute cannot be made to support the dire philosophical charges usually brought against Kuhn. Admitting that is not (of course) equivalent to supporting Kuhn's specific thesis about paradigm shifts (or his own version of incommensurabilism or his weak defense of his own thesis). That must be remembered. But it is still the first step in the argument.

2

Let me put this more manageably. If you hold that nature is a closed system, meaning by that that science progressively grasps the single inclusive order that *is* the independent world (call that 'objectivism' or 'Cartesianism'), the incommensurabilism thesis will be ruled out at once – *not* for the reasons Davidson suggests, namely, that *completely* independent (read: *not* in any way or degree intertranslatable) conceptual schemes cannot even be grasped (which is true enough but irrelevant) or that *partially* independent (different) conceptual schemes risk the anomaly of being independent *parts* that are too independent to be parts of any one scheme (which is surely a dreadful confusion).[18]

The reason is simpler and more telling than Davidson imagines: for, *if* the independent world can be discerned as it is independently, without interpretive props of any kind, then it *is* a system of sorts, and then it *is* coherent *qua* that system, and then there would cease to be any Kuhnian-like role for epistemological incommensurabilities (in effect, incommensurabilism proper) to play – that is, *at* the point at which we do grasp the way the world is! But *if* that prospect – which (as I say) is now often called 'objectivism'[19] – were indefensible or unconvincing or inaccessible, then whatever we construed *as* the independent world (on our best evidence) would always be subject to interpretive intermediaries that we could never altogether eliminate.

Then, although we might indeed hope to grasp the inclusive system of the whole of nature, we could never rule out a plurality of partial, competing accounts of the independent world – which is all that would be needed initially in order to allow Kuhnian-like incommensurabilities to flower, although that alone would hardly establish Kuhn's (or Feyerabend's) thesis. (Bear in mind that more innocuous or elementary incommensurabilities than Kuhn's would always obtain.)

Our picture of the objective world would then be a *construction*, and because of that, large-scale incommensurabilities could never be ruled out. That would *not* be because the world was constructed or 'mind-dependent' in any purely ontic sense (in effect, idealism's characteristic claim), but because whatever, on epistemic grounds, we rightly judged to be *ontically independent* would count as such *only on* the sufferance of its being *epistemically dependent* (that is, confirmed). This condition cannot be satisfied except in the constructivist way; that is the price of rejecting Cartesianism. There's the pivot of the matter.

What it shows – perhaps surprisingly – is that the outright rejection of incommensurabilism, as unintelligible or paradoxical or self-defeating or incoherent, is no more than a vestigial corollary of Cartesianism itself, and that constructivism (which opposes Cartesianism and which cannot fail to tolerate the possibility of incommensurabilism) is *not* necessarily a form of idealism at all, but only a denial (in effect, a denial of the closed Cartesian universe) of the belief *that* metaphysical and epistemological (or, semantic or logical or psychological) questions *can* be disjunctively pursued.

The bare admissibility of the incommensurabilism thesis is bound to the acceptance of something akin to the following triad: (i) the inseparability of epistemological and metaphysical questions regarding the objective world; (ii) the indefensibility of objectivism, and (iii) the epistemic inescapability of contingent interpretive intermediaries. The first item is the most general lesson implicated in the defeat of Cartesianism, laid down by the contributions of Kant and Hegel, without our being forced to adopt their particular doctrines; the second concurs with Kuhn's rejection of epistemic neutrality in the correspondentist sense, the natural mate of the first; and the third is the inference that, on the strength of the first two assumptions, no viable realism can be more than a constructive and historicized realism (a constructivism), not for that reason alone a form of idealism (which would be a perverse reading of the first assumption).

This does not quite tell us what to count as a viable specimen of incommensurabilism in the Kuhnian sense, but it permits us to retrieve, without controversy, ordinary forms of conceptual incommensurability on which the other builds, namely, whatever admits the absence of a common metric (expressed in integers) by which, say, all the sides of a right-angled triangle would, if commensurable, be able to be measured by the same metric – hence, by extension, the absence of any suitably analogous conceptual instrument that behaved not unlike a metric, in the way of straightforward comparison, subsumption, generalization, or the like (as in speaking of the different visual appearances of the sky, for instance). The important lesson here is, first, that incommensurability (among attributes) is widespread; second, has nothing, as such, to do with being unintelligible, incomparable, untranslatable, or even eliminable, and third, is so bland a doctrine that its bare admission is even reconcilable with nature's being a closed system.

Let me put the last remark a little more perspicuously: incommensurabilities associated with the Pythagorean case are so simple that they have almost no bearing on the larger issues theorists like Davidson and Putnam bring against Kuhn and Feyerabend; correspondingly, Kuhn's and Feyerabend's objections to treating science's picture of nature as a progressively completed account of a single closed system make room for deeper incommensurabilities than those envisaged by the simple case. These two observations count as a gain, because they confirm that the *a priori* rejection of large-scale epistemological incommensurabilities is a direct consequence of advocating some version of the objectivism Kuhn and Feyerabend contest. It's the failure to see that Kuhnian-like incommensurabilism hangs on a deeper puzzle than do Pythagorian and similar incommensurabilities that befuddles the debate.

Let me put another argumentative nail in place to suggest how to move from the elementary cases to the Kuhnian case. If, say, the lengths of the sides of a right-angled triangle are incommensurable with that triangle's hypotenuse, then so, too, are the sets of different colors and different lengths of textile swatches *vis-à-vis* one another (that is, colors and lengths). The lengths of the right-angled triangle's sides plainly remain comparable – the hypotenuse being longer than the other sides – though they remain incommensurable in the technical sense; and the color of any swatch will be incommensurable with its length, and vice versa, in the obvious sense that lengths do not fall under the category 'color' or colors under the category 'length.' Although, to be sure, the *categories* 'color' and 'length' can be compared as distinct categories that are not subsumed under one another, that range over the same particulars (swatches of fabric), and may belong coherently to a larger descriptive scheme. All such incommensurabilities will be marked within a linguistic practice apt for doing so! So it is reasonably clear that to think (with Davidson) of conceptual schemes that are utterly or 'partially' unrelated to one another – 'conceptually incommensurable' only in *that* Pickwickian sense – is to confound utterly *any* ordinary sense in which incommensurables are admitted. Finally, if all this is so, then incommensurable concepts (as distinct from the incommensurabilist doctrine) must be ubiquitous, benign, paraphraseable, not necessarily encumbered by the historical drift of paradigms or the historicity of thought itself.

This is not yet the place to pit Kuhn and Davidson against one another. But you will find it helpful to remember that, when he attacks Kuhn's and Feyerabend's views on incommensurability, Davidson unaccountably conflates Kuhn's well-known rejection of neutrality with the 'failure of intertranslatability' and with the bare admission of conceptual incommensurability.[20] This is a multiple blunder: because incommensurables *are* translatable *qua* incommensurables; because their being translatable does not presuppose or entail any neutral stance or language; because the so-called untranslatability of incommensurables signifies no more than that translation cannot overtake incommensurability itself; and because, on Kuhn's own thesis, incommensurability does not imply *any* failure of language at all.

Here is what Davidson says – famously:

> The failure of intertranslatability is a necessary condition for difference of conceptual schemes; the common relation to experience of the evidence is what is supposed to help us make sense of the claim that it is languages or schemes that are under consideration when translation fails. It is essential to this idea that there be something neutral and common that lies outside all schemes. That common something cannot, of course, be the *subject matter* of contrasting languages, or translation would be possible.[21]

Obviously, Davidson is committed to some version of neutrality, *not Kuhn*, for he (Davidson) holds that if there were incommensurable conceptual schemes (which he opposes) then translation would necessarily fail. Davidson nowhere defends the idea. The best one can say on Davidson's behalf is that he may wish to fall back to objectivism. But Kuhn's cases do not require 'independent' conceptual schemes (in Davidson's sense): that possibility is a complete red herring. Davidson neglects to consider the garden-variety specimens already sampled. But *they* hold the master key to the entire puzzle.

I must enter a very large stage whisper here, because I have not officially acknowledged that Davidson's principal target in his attack on Kuhn and

Feyerabend is really his own mentor, Quine, as the author of the 'indeterminacy of translation' thesis in *Word and Object* and the eventually articulated account of the epistemic 'neutrality' of the perceptions represented by the 'holophrastic sentences' of Quine's *Pursuit of Truth*.[22]

Actually, once you realize that Quine *is* the target of the charge formulated in Davidson's 'third dogma' of empiricism, you see at once the coded meaning of Davidson's and Putnam's own dispute, the obvious (but still indefensible) deformation of Kuhn's and Feyerabend's accounts of science, and the utter irrelevance of Davidson's opportunistic use of the scheme/content disjunction in attacking the important epistemological puzzle Kuhn and Feyerabend have unearthed in the name of incommensurabilism. But here, it is the substantive issue that I mean to salvage. I leave for another occasion the matter of correcting a philosophical injustice. Also, I think it fair to say on Quine's behalf that Davidson offers not the slightest reason to suppose that Quine would be willing to construe his own admission of incommensurability (in the 'indeterminacy of translation' issue) along the lines Davidson has laid out.

You must bear in mind that, for Kuhn, Galileo (Kuhn's exemplar in the matter of mastering epistemological incommensurables) is, in effect, a bilingual completely at home among well-articulated, relatively large-scale explanatory systems that *are* epistemologically incommensurable. Kuhn admits no massive 'failures' of language; no neutral world, and above all, no closed encompassing language or conceptual scheme of any kind. These features are imputed, whole cloth, in Davidson's account, to Kuhn. Objectivism (or Cartesianism) requires that we *read* Kuhn in this dismissive way. But if the foregoing analysis is reasonably correct, then incommensurabilities are always *local* (no matter how extensive or ramified), always *relationally defined* (relative to some favored metric or conceptual analog); hence always capable of being *mastered* both monolingually (so to say) and bilingually.

Furthermore, *if* Kuhnian-like incommensurabilities accord with our most elementary specimens, then we should admit that *they* (the garden-variety cases) neither presuppose nor require any neutral, inclusive, single, conceptually closed scheme or language; or that *their* detection presupposes or entails the violation of any neutral scheme or language. All viable languages admit bilinguals and their conceptually competent counterparts: as in mastering dialects, idiolects, classificatory schemes, explanatory theories. Hence, *if* Priestley and Lavoisier lived in 'different worlds,' as Kuhn notoriously affirms,[23] then *they could* still understand one another. *They* would never need to share a neutral language: in fact, there is in principle no difference between the interlinguistic and the intralinguistic.

3

Now, what I claim is this: *all* the forms of incommensurability are variations or complexifications of the two kinds of cases I mentioned just above, namely, the Pythagorean case and that of different color swatches of different sizes. The most quarrelsome case, which I am keeping in the wings for the moment, is of course Kuhn's and Feyerabend's, occupied with the historical drift and evolution of thought. I take the latter case to depend on the separate fate of objectivism, but I insist that its bare coherence requires that it be brought into accord, conceptually, with the simpler cases (or with others comparable to them).

Here, a fresh advantage suggests itself. The simplest cases are clearly occupied with predicative distinctions. Could there be, then, a form of incommensurability that holds even among normally compared predicates – colors or spatial relations – taken separately? Of course there can! There are some convincing comparative studies that help to extend the argument – involving the prelinguistic perceptual training of infants growing up in different languages – that yield obviously incommensurable distinctions that are legibly codified nevertheless.[24] The matter is complicated, to be sure, by the learned/innate disputes made famous by Noam Chomsky, as well as by disputes about the role of prelinguistic social learning. But I waive such questions here, except in the sense of leaving them deliberately unresolved – though not in any tendentious way. It is certainly possible that Chomsky's master notion of an innate, underlying universal grammar is mistaken: in the sense, first, that syntax may be inseparable from semantics, and second, that for any such grammar (UG), a given natural language may violate its putative constraints. In fact, Chomsky has recently signaled his decision to abandon the model of UG for which he is so well known.[25]

In any case, it has been shown that the spatial use of the prepositions *in*, *on* and *under* (in English) relies on very distinctive 'containment'/ 'noncontainment' and 'support' features (mingling semantic and syntactic considerations) critical for the systematic description of objects ('the cup is on the table,' 'the picture is on the wall,' 'the leaf is on the tree,' 'the band-aid is on the shoulder,' 'the napkin-ring is on the napkin'), and that the use of cognate German prepositions (*auf*, *an*, *um*) is more sensitive than English as 'to whether a relationship of contact between two objects involves a relatively horizontal surface [*auf*], a vertical or otherwise non-horizontal surface or contact point [*an*], or encirclement [*um*].' The corresponding distinction in Dutch between *op* and *aan*, it is said, 'has less to do with orientation than with *method of attachment*.' Spanish apparently bypasses the 'on–in' continuum by using *en* for both ('an apple *en* a bowl,' 'a cup *en* a table'). Several Mixtec (middle-American Indian) languages, which lack prepositions and other locative markers, apparently express something like the relevant spatial relations by way of 'an extended and systematic body-part metaphor' (that separates, in addition, animal and human body-part analogies), in some cases (and in some cases not) requiring contact and support, or adjacency, or other configurations.[26]

All these regularities seem to be equally easily learned, and yet, of course, the classifications in each language need not be strictly 'commensurable' (though they remain comparable) with those of the classificatory schemes of other languages (in different degrees and in different respects) and in a way that might even be replaced by a more inclusive commensurable scheme. Certainly, the reported features of the Mixtec languages suggest how widespread the incommensurability may be, even with regard to the most elementary categories of description. For example, it is said to be particularly difficult to learn the languages of the peoples of the South Pacific islands, apparently in large part because their categories embody the exotic practices and beliefs of notably local societies: under such circumstances, it is very likely that predicative incommensurabilities are bound to arise, incommensurabilities that would be difficult to fathom; in fact, it is likely that such languages may not exhibit the high degree of grammaticalization (finding favored universality) that characterizes the usual linguistic specimens of Western theories of language.

Incommensurabilities of these sorts remain, nevertheless, both real and benign, and are both paraphraseable and replaceable in principle. The important point to remember is that, if (as I would insist) we eschew all appeal to Platonic Forms or natural essences, then the commensurability of categories between natural languages will be hostage to the tolerances of what to count as the perceived similarities or acceptable extensions of given general predicates (whether construed intralinguistically or interlinguistically). But that means that incommensurabilities of this third sort lurk everywhere, cannot be eradicated, and produce no particularly paradoxical effects on our fluency (though we must be careful).

It takes but a step to realize that we can afford to allow this sort of phenomenon to be subject, further, to historical drift, as a result of actual collective usage – which even Wittgenstein admits (in a very thin way), though he is no friend of historicism or historicity.[27] If you concede that the paradigm of perception – the human case – is 'enlanguaged,' cannot be segregated between a non-linguistic sensory element that is first 'given' and then linguistically reported (which would, of course, invoke Cartesianism again),[28] then the bare admission of incommensurabilities at the reporting level (subject, further, to historical drift) would effectively confirm the coherence of epistemological incommensurabilities of Kuhn's and Feyerabend's sorts. That would clinch the issue.

I have, at the very least, provided four grades of incommensurability that are clearly more fundamental than the grand sort of conceptual incommensurability Kuhn and Feyerabend feature and Davidson has so effectively (but misguidedly) attacked. They seem to me to form a family of the following sort. First of all, they are clearly intelligible, describable, ubiquitous, open to fluent natural-language use. Second, their being such presupposes a more inclusive linguistic practice in which incommensurables may actually be sorted, though there is no reason to suppose that the encompassing language forms a single or inclusive classificatory system or can be counted on to be neutral in the objectivist's sense. Third, the kinds of incommensurability in question are all predicative in nature, hence subject to the intractable difficulties known to affect the local problem of discernible predicative generality.[29] And fourth, the three sorts of elementary incommensurability (as we may call them) are all, on empirical grounds, subject to historical change and divergence, without any loss of natural-language fluency and without overcoming the constant threat (if it be a threat) of continuing conceptual incommensurability.

I see no way to gainsay these distinctions. They confirm the general finding that the analysis of incommensurability should be separated, without prejudice of any kind, from the analysis of Kuhn's and Feyerabend's cases, except for the important conjecture that the latter cases (Kuhn's and Feyerabend's) may be shown to be extreme instances of, or to depend on, the grades of incommensurability now collected, and except for the argument that merely historicizing thought normally will not make any of our specimens paradoxical or incoherent.

If all this be conceded, then a decisive part of Kuhn's and Feyerabend's argument will have been shown to be coherent, for the historicized drift of categories will be seen to be a contingency affecting commensurable and incommensurable predicates alike. If the historicity of thought is not inherently paradoxical, then, so far at least, the Kuhnian case will be as benign as the other sorts of incommensurability already collected. I am not claiming (here) that Kuhn's and Feyerabend's deeper incommensurabilist arguments actually hold (though I am prepared to defend them), only that they cannot be faulted simply because they involve incommensurable predicates and entail the historicity of thinking.

It's worth remarking that the familiar critics of Kuhnian incommensurability do not normally attack on grounds of historicity's incoherence; they argue, rather, as Davidson does, on grounds of the supposed formal incoherence of conceptual incommensurability, which, as I have already suggested, presupposes the modal necessity of something like the Cartesian or objectivist position (however pared-down it may be).

In any case, it is a striking fact that Davidson almost never pursues epistemological questions in depth, or questions of scientific perception; so his animus against Kuhn and Feyerabend is almost certainly based on abstract philosophical objections against admitting any pertinent puzzles that would oblige him to turn to epistemology itself! This helps to explain how his worries about Quine's 'indeterminacy of translation' thesis may have led him to an easy (but entirely indefensible) conflating of Kuhn's and Quine's puzzles. They are altogether different, in the sense that, although Quine's view[30] does indeed make room for incommensurability (as in the parsing of 'Gavagai!'), Quine's 'translations' depend on some supposed common perception (represented by 'holist' or 'holophrastic' sentences); whereas Kuhn's conception is more daring in that the incommensurability he considers (Priestley and Lavoisier's, for instance) directly affects what would otherwise be a common perceptual ground. Admitting that much (in Kuhn's case) has *nothing* to do with violating Davidson's injunction against invoking the scheme/content dogma.

If, further, all the forms of cognitive privilege (that might have buttressed our claims of neutrality and conceptual closure) fail – and there is almost no one today (practicing in the analytic manner) who would be willing to champion foundationalism, apodicticity, conceptual 'totalizing' or cognitive privilege – then it is a foregone conclusion that historicity cannot be ruled out as impossible or improbable. There seems to be *no* demonstration of the supposed incoherence of historicity that does not rest (by way of a *petitio*) on the claim that the basic categories of thought must be invariant and known and universally present in all natural languages. Anything less would instantly lose the modal force of the intended objection.[31] Reflect, for instance, on the implications of Davidson's unsatisfactory conjecture that 'most of our beliefs are true.'[32] Davidson must mean at the very least that the beliefs of different peoples must, even if construed from our own perspective, be predominantly true. But how does he know that? And how would it be possible if epistemic incommensurabilities and historicized divergences obtained? He nowhere addresses the question.

If predicative generality is labile in the logical respect remarked – that is, if it is linguistically encumbered, never more than consensually grounded, always subject to historical drift – then invariance cannot be more than a construction of convenience. You see, therefore, how difficult it is to recover the Cartesian argument (the conceptual closure argument). Alternatively, you see the sense in which the objectivist's attack on incommensurabilism harbors some secret form of Platonism.

4

This brings me, finally, to Kuhn's and Feyerabend's well-known claim. We may now risk addressing their particular texts directly, without fear of losing the essential thread. I've now laid a proper foundation for separating the bare predicative issue from the inevitable complications of Kuhn's and Feyerabend's respective theories of science, that is, their incommensurabilism. By the same token, we may examine Davidson's analysis more closely. For, certainly, Davidson's reading of Kuhn and Feyerabend has dominated the analytic camp's reception of their best-known texts. In fact, Davidson cites them and provides what many regard as the strongest reasons for supposing *that there are no 'basic'* (Davidson's word) *conceptual incommensurabilities*: apparently, there can't be any!

I'm afraid I shall try your patience here: the argument cannot avoid a fussy textual turn. I enter the quarrel in order to ensure a final reckoning regarding the connection between Kuhn's and Feyerabend's views and the elementary forms of incommensurability already laid out.

Here, then, is what Davidson cites from Feyerabend:

> Our argument against meaning invariance [this is Feyerabend speaking] is simple and clear. It proceeds from the fact that usually some of the principles involved in the determination of the meanings of older theories or points of view are inconsistent with the new … theories. It points out that it is natural to resolve this contradiction by eliminating the troublesome … older principles, and to replace them by principles, or theorems, of a new … theory. And it concludes by showing that such a procedure will also lead to the elimination of the old meanings.[33]

If I understand Feyerabend's remarks and Davidson's analysis aright, Davidson acknowledges the case Feyerabend describes, but asks us to consider:

> only whether, if such changes were to take place, we should be justified in calling them alterations in the basic conceptual apparatus … What is clear [Davidson says] is that retention of some or all of the old vocabulary in itself

> provides no basis for judging the new scheme to be the same as, or different from, the old. So what [he adds] sounded at first like a thrilling discovery – that the truth is relative to a conceptual scheme – has not so far been shown to be anything more than the pedestrian and familiar fact that the truth of a sentence is relative to (among other things) the language to which it belongs. Instead of living in different worlds, Kuhn's (probably then, also, Feyerabend's) scientists may, like those who need Webster's dictionary, be only words apart.[34]

Bear in mind that Davidson advises us (just before offering his comments) that *he* 'think[s] our actual scheme and language are best understood as extensional and materialist' and that, 'if we were to follow this advice, ... [neither] science [n]or understanding would be advanced, though possibly morals would'[35] – by Feyerabend's case. Of course! The implication is that there *is* a 'basic conceptual apparatus' in natural language and human thought and that, admitting *that* (Davidson's objectivism), the incommensurabilities Kuhn and Feyerabend feature cannot possibly serve to demarcate different conceptual schemes *in the 'basic' sense* and cannot lead to the discovery (which Davidson assumes *they* intend) that, say, 'truth is relative to a conceptual scheme.' I put it to you that Davidson's counter-move is an *obiter dictum*, an undemonstrated and utterly indemonstrable 'first' principle if meant to be categorically affirmed. Furthermore, it actually concedes the possibility of incommensurabilities in the benign sense already supplied. But, apart from all that, it shows very clearly that Davidson harbors a privileged view of the conditions of cognition (of reality) that his usual work (as in 'A Coherence Theory') refuses to allow – on pain of generating an insuperable skepticism.

I admit this is a harsh judgment against Davidson. But it is a fair one. Let me offer two considerations straight off. For one, Feyerabend *never* invokes a 'basic conceptual apparatus' that could be explicitly formulated or would be coextensive with the whole of a language or the entire closed system of nature. What Davidson cites from Feyerabend concerns changes only in ideal explanatory theories or congruent changes in the meaning of both descriptive and explanatory terms – made necessary under the pressure of finding that some original theory *cannot* explain (possibly cannot even make sense of) a phenomenon that rightly calls for explanation. For instance, Feyerabend notes, more or less in agreement with Kuhn, that the inertial motion Galileo and Descartes consider 'does not make sense, nor can it be formulated' within 'the Aristotelian theory of change.'[36] Failure to be formulable in the terms of one theory despite being formulable in those of another, failure to capture what is descriptively intelligible in observed events by means of a theory thought capable of explaining such events, failure to eliminate inconsistencies between two theories said to be (in a formal sense) separately adequate as explanations of the same events, count as specimen instances of what Feyerabend treats as incommensurabilities.

Furthermore, in the paper cited, Feyerabend *nowhere* draws conclusions remotely like Davidson's, that is, that 'truth is relative to a conceptual scheme.' That thesis is incoherent on its face (as Davidson very well knows), *if*, by 'relative to a conceptual scheme,' Davidson intends something close to what Myles Burnyeat proposes as a reading of Socrates' account of Protagorean relativism (in the *Theaetetus*): namely, that 'true' means 'true-for-*x*' (where *x* signifies one or another speaker, or even one or another conceptual scheme) – what I have already termed 'relationalism' in reviewing Putnam's account.[37] Otherwise, at best, Davidson's point is trivial.

Certainly, Davidson opposes Feyerabend on grounds of a threatening and incoherent relativism. Although, of course, the mere admissibility of incommensurabilism entails assigning the theory of truth an explanatory role. If so, then even Davidson's account of canonical science cannot justify denying truth an explanatory or legitimative role – without, however, our insisting that, as a result, truth must have a criterial role as well. No. That last claim would be Cartesianism pure and simple.

But now, it turns out that Davidson's insistence that conceptual incommensurability is intrinsically incoherent is also Cartesian (in the sense supplied); for the objection to incommensurable conceptual schemes depends on the supposed fact that they cannot be 'alterations in the basic conceptual apparatus' of our entire truth-seeking inquiries: they would need to be *compared* with the categories of our former scheme, and then, *putative within the resources of our 'entire' language*, we would still not be able to say whether they were 'the same as, or different from, the old.' 'The "old" *what*?' you ask: the 'part' of the 'whole' language we are using, or the concepts there invoked, that harbor 'the basic conceptual apparatus' we are alleged always to employ? The matter remains a complete mystery: Davidson never tells us what that apparatus includes. In any case, the objection is obviously irrelevant, because, as I have already shown, conceptual incommensurabilities belong to natural languages as such and are entirely open to translation and comparison. The only sense in which Davidson could possibly hold to his extreme finding would require that what he calls 'the basic conceptual apparatus' *is*, for reasons bordering on necessities *de cogitatione*, necessary and inviolable. He's much too adroit to admit any such conviction, but it lurks in the neighborhood nevertheless.[38]

In effect, Davidson mismanages Feyerabend's entire argument: because he discounts Feyerabend's claim on grounds that could never have had a place in the latter's philosophy, and because he himself disregards the *local* reading of incommensurability (which he implicitly concedes). The whole point of Feyerabend's brief is to show that a purely deductive model of explanation and the corollary notion that changes in explanatory theory do not (or never) entail changes in the meaning of descriptive and theoretical terms goes entirely against the history of science. The intended lesson is that the failure of these two doctrines is sealed by the actual epistemological incommensurabilities that have occurred at critical moments in the history of the physical sciences that, by their appearance, preclude a closed system of nature and 'a basic conceptual apparatus.'[39] Feyerabend is dead right in what he offers in evidence against such theorists as (Rudolf) Carnap and (Carl) Hempel and (Ernest) Nagel and (Nils) Bohr. Davidson's charge is a complete non-starter. As a consequence, Davidson never actually addresses the logic of incommensurability, though that may be hard to believe. I see no way to avoid the verdict.

5

You realize here that Feyerabend's challenge calls into question any realism that proceeds without interpretive intermediaries, that presumes on the avoidability of constructivism, that embraces Cartesianism, that resists *ab initio* the very idea of a viable relativism.

But if you accept this way of answering Davidson, you must concede as well that Feyerabend's kind of case (which is also Kuhn's) does not alter in any fundamental way the table of the four grades of incommensurability already given. What we see from Feyerabend's account is that if incommensurabilities arise in the elementary settings mentioned, then they are bound to arise in disputes about much larger explanatory theories. Incommensurabilities are in a sense always 'local,' always relativized to particular linguistic resources, intelligible as such, empirically motivated, though not captured in any narrowly empirical (or empiricist) way, never keyed to translatability *überhaupt* or such as to presuppose completely separable conceptual schemes, but only such (schemes) as result from normal changes in scientific practice and the historicity of thought. The idea of large-scale incommensurabilities that remain 'local' does not entail any paradox or important novelty regarding their analysis, though it does affect the bearing of history on the theory of explanation – and of knowledge and understanding and of truth and meaning.

The puzzle of the historicity of thought is at least as important as the puzzle of local incommensurabilities.

Consider, now, what Davidson cites from Kuhn:

> Philosophers [this is Kuhn speaking] have now abandoned hope of [finding a pure sense-datum language], but many of them continue to assume that theories can be compared by recourse to a basic vocabulary consisting entirely of words which are attached to nature in ways that are unproblematic and, to the extent necessary, independent of theory … Feyerabend and I have argued at length that no such vocabulary is available. [This, at least implicitly against Pierre Duhem and in agreement with Quine.] In the transition from one theory to the next words change their meanings or conditions of applicability in subtle ways. Though most of the same signs are used before and after a revolution – e.g. force, mass, element, compound, cell – the way in which some of them attach to nature has somehow changed. Successive theories are thus, we say, incommensurable.[40]

Davidson's instant comment runs as follows: '"Incommensurable" is, of course, Kuhn's and Feyerabend's word for "not intertranslatable". The neutral content waiting to be organized is supplied by nature.'[41] This is as strange as the remarks made about Feyerabend. For one thing, Kuhn is clear that *he* does not oppose translation *anywhere*. Where, in the context of incommensurability, he warns about translation – in fact he makes the point immediately before the passage Davidson cites (which is from the essay 'Reflections on My Critics') – Kuhn confines the sense of his remark to what he calls 'the point-by-point comparison of two successive theories [which] demands [he says] a language into which at least the empirical consequences of both can be translated without loss or change.' Many have sought such a language, he continues, which they suppose is 'neutral' (for instance, the language of 'pure sensation-reports') or possesses 'a 'universal character' which would display all languages for expressing them as one.' Then comes the line that Davidson cites: 'Feyerabend and I have argued at length that no such vocabulary is available.'[42] Nothing could be plainer. Kuhn clearly holds that both he and Feyerabend oppose *any* 'neutral content waiting to be organized' or *any* neutral language for descriptive and comparative use. There is, he says, no 'basic vocabulary'; *a fortiori*, there is no 'basic conceptual apparatus.' Davidson misleads us here.

Note that this demurrer does *not* mean that incommensurable statements are not translatable, or intertranslatable. Where they are translated ('point-by-point'), they require 'compromise' (Kuhn says), precisely because of the problematic intertranslatability of sentences from different languages. Notice, please, that Kuhn actually adds, a few paragraphs later:

> To me at least, what the existence of translation suggests is that recourse is available to scientists who hold incommensurable theories. That recourse need not, however, be to full restoration in a neutral language of even the theories' consequences.[43]

Davidson ignores the point. Kuhn goes on to enlist Quine's famous 'indeterminacy of translation' thesis, chides his own critics for being too vague about translation 'point-by-point,' acknowledges the comparability of incommensurable theories throughout, and emphasizes the very sensible view that whatever will finally count as an adequate translation will depend on interests that are bound to require and allow compromise.[44]

Davidson unaccountably foists on us the misleading equation of 'incommensurability' and failure of 'intertranslatability'; and, where Kuhn expressly remarks that he and Feyerabend reject neutrality, Davidson assigns Kuhn the gratuitous view that 'nature' somehow supplies the 'neutral content' that our incommensurable schemes will thereupon 'organize.' There is no warrant for either of these glosses, and there is no 'neutral content' to be had. Davidson constructs, and duly defeats, a so-called 'third dogma' of empiricism (beyond Quine) – the 'dualism of scheme and content,' which he wrongly assigns to Kuhn and Feyerabend[45] – though there is almost no one who ever held the 'dogma' Davidson defeats. (He apparently draws it, at least in part, from Quine.) The only plausible candidate is Benjamin Whorf, but Whorf cannot be compared with Kuhn and Feyerabend on the history of science.[46]

There's more that could be said about Davidson's unusual explication, but it would take us too far afield from the incommensurability issue. Suffice it to say that Davidson's notion of 'conceptual relativism' relies on the paradoxical 'relationalism' already mentioned, wrongly characterizes Kuhn's account of incommensurability, and then dismisses the fiction he has invented.

I am bound to say in all fairness that Kuhn defends the incommensurabilism issue rather poorly, though he collects the important specimens of explanatory incommensurability. What needs to be recovered is the important question of what we should understand by an objective science under the condition of the historicity of thought. But I see no reason to take up the defense of Kuhn's specific picture of science here – especially where Kuhn is prepared to retreat to the sociology of science and Feyerabend, to endorse his well-known brand of anarchism.[47]

Two further caveats are in order. For one thing, there is no warrant whatsoever for introducing – as a proper instance of incommensurability or its principal form – Davidson's case of 'completely' or 'partially' independent conceptual schemes (in the sense he intends). Where admitted, it is totally inoperative, marks no more than the limit of bare intelligibility. For another, granting that, there is no ground for any additional disjunction between so-called 'incommensurability of *meaning*' and 'incommensurability of *standards*.' On this question, Ronald Giere has the following to say:

> In the process of creating a normal science research tradition, a scientific community develops 'norms,' 'rules,' or 'standards.' These help determine, among other things, what counts as a significant problem and what counts as a solution. But Kuhn's doctrine of the priority of paradigms (that is, exemplars) holds that the standards are determined by the exemplars, not the exemplars by the standards. It follows that a new solution to a problem, one appealing to a new exemplar and thus violating the standards of the older research tradition, could not be recognized as a solution by that tradition.[48]

But that's a misunderstanding and yields too far in Davidson's direction. At best, it trades on an equivocation: 'not recognized' sometimes means no more than 'not officially countenanced' though intelligible enough to those who work within the tradition – Lavoisier and Priestley, say, within the tradition of eighteenth-century European chemistry. Giere simply ignores the principal question.

In the relevant sense, pertinently motivated changes of standard may well entail changes of meaning. It's not the sheer priority of recognizing the need for a new standard – which may then usher in a solution that 'could not be a solution within a [given] tradition' – that is decisive; it's that what *is* a promising or adequate solution *cannot, for deeper epistemological and methodological reasons*, be captured *commensurably* by the terms favored in the canon of the older tradition. It *can* be 'recognized' by that tradition. It has nothing to do with whether the scientists of the older tradition are able or unable to *understand* the solution of the new 'paradigm.' It's merely that the new solution would not have been 'countenanced' by the older one, given its epistemological stance. 'A solution to the problem of atomic spectra,' Giere offers, 'that invokes quantized values for the exchange of energy, for example, could not be a solution within a classical tradition that presumed continuity. In recognizing a new solution as indeed a solution, therefore, scientists are creating a new tradition, and thus new standards. But their prior decision to accept the new solution is then not governed by any standards at all. It is a truly revolutionary act in the classical, political sense.'[49]

Perhaps. But it *is* governed by the conditions of intelligibility that the champions of both traditions effectively draw on and that Kuhn rightly commits himself to in the remarks I have cited. In the sense at stake, incommensurable changes of meaning may (often do) implicate incommensurable standards; and incommensurable standards may (often do) entail incommensurable meaning-changes. That's all that's needed. Giere cannot mean – or does he? – that, in the puzzle cases Kuhn collects, the issue never turns on weighing *epistemological incommensurabilities* (that is, the incommensurabilist's specific claim). If he denies that there are any such incommensurabilities, he has yet to provide the argument; and if he admits them, then his own argument misses the mark.[50]

6

It needs to be said, finally, that the key to Kuhn's and Feyerabend's distinction rests with their explicit repudiation of neutrality (*not* objectivity) and their implied rejection of any discernible conceptual closure regarding the physical world that disallows the inventive possibilities of science and epistemology that supply the ground for challenging any would-be closure. Neutrality makes no sense except in terms of an epistemology able to demonstrate the undistorted correspondence between the independent world and our would-be science. But that cannot be shown, on pain of all the familiar paradoxes of the 'Cartesianism' that runs from Descartes through Kant.

The very idea of the conceptual plenitude of our natural-language resources (Davidson's thesis) signifies that no creative science could possibly generate concepts incommensurable with our 'basic' vocabulary – for if it did, so the argument goes, the resultant science would be unintelligible. But then the argument presumes, indemonstrably, that the conceptual resources we *now* possess (in effect, possess at every stage of our continuous history) *is* constitutive in some robust way in the conceptual or causal closure of the world. That closure (which we need never actually master) apparently ensures the ultimate neutrality of our evolving science; furthermore, the continuing intelligibility of science itself, even under radical change, is said (as by Davidson) to confirm the absence of any 'basic' incommensurability. But remember: we can never actually formulate what the 'basic' vocabulary includes. To be able to do that, we should already possess a clear sense of the closure of the world under the conditions of causal explanation.

That would be a very tidy theory, you must admit. But it suffers from two serious flaws at least. For one, it has no relevance whatsoever for appraising either first-order theories like Priestley's and Lavoisier's or second-order theories like Kuhn's and Feyerabend's regarding the incommensurability of competing first-order theories. The reason is plain. *If* the theories in question *are* intelligible and comparable and paraphraseable, then, *on Davidson's account*, they *are* commensurable with our 'basic' concepts even if they are locally incommensurable with one another! (And, of course, even if – or, conveniently enough, because – they are unknown or unknowable.) That they are commensurable in Davidson's sense is little more than a trivial consequence of their being intelligible. But then, Kuhn's and Feyerabend's incommensurabilities *are* commensurable with our basic vocabulary in the same empty sense in which commensurable but competing first-order theories are said to be. Second, Davidson provides no further account of incommensurability that demonstrates its incoherence.

I am *not* claiming that Davidson holds that there is *one* conceptual scheme that we actually possess that collects the basic vocabulary postulated. No, not at all. What I suggest rather is that he believes that whatever is intelligible in the conceptual history of science, including differences in competing theories, *entails* a 'common' network of concepts answering to the objective work of an ongoing science. In that sense, there is a 'common' vocabulary that we *must* share; but it cannot be assigned a determinate number or explicit content, since that would signify that another conceptual scheme *might* be formulated in opposition to it – and that, Davidson thinks, is impossible because it is incoherent.

Thus he says, at the very outset of 'On the Very Idea …':

> Even those thinkers who are certain there is only one conceptual scheme are in the sway of the scheme concept: even monotheists have religion. And when someone sets out to describe 'our conceptual scheme,' his homey task assumes, if we take him literally, that there might be rival systems.[51]

This is the essential move, Davidson believes, to dismantling 'conceptual relativism.' He does not mean that there is only one system: he means that there cannot be 'rival' systems; and if there cannot be rival systems, then the concepts we use must accord with what is 'basic' to the 'neutrality' or objectivity of *our* science (such as it is) *and* signifies the only kind of neutrality that makes sense. For Davidson, it appears to signify a kind of necessity that stands confirmed without any further determinacy – without even the possibility of fathoming its conceptual content.

But consider Davidson's deeper argument:

> The dominant metaphor of conceptual relativism, [he says,] that of differing points of view, seems to betray an underlying paradox. Different points of view make sense, but only if there is a common co-ordinate system on which to plot them; yet the existence of a common system belies the claim of dramatic incommensurability.[52]

Now, that's a fatal *non sequitur*. For one thing, neither Kuhn nor Feyerabend is committed to supporting anything like Davidson's 'dramatic incommensurability'; and, for another, our ability to compare two different points of view does not entail that 'there is a common co-ordinate system' that we share in making that comparison, *if* that is supposed to override the incommensurability of Kuhn's and Feyerabend's examples. It's true that our language permits us to *compare incommensurable points of view*; Kuhn and Feyerabend concede that much: witness Galileo. But comparability does not eliminate Kuhnian incommensurability, and does not drive it to the limit of unintelligibility. Furthermore, there is no sense in claiming a 'common co-ordinate system' if we cannot state what that system is; and if, *per impossibile*, we *can* specify the details of that system, then *there is one conceptual scheme* after all – and that, Davidson insists, makes no sense.

These are petty quarrels perhaps, but they are also tactical maneuvers in the service of a larger strategy that means to keep alive the double theme Kuhn and Feyerabend espouse: that is, the rejection of neutrality in the guise of an objectivity free of the constituting function of our conceptual schemes, and the rejection of the idea that nature forms a closed system of such a sort that the concepts of an objective science must match that system, and matching it, ensure neutrality thereby.

Nevertheless, neutrality remains a Cartesian vestige that cannot escape its original paradox. Davidson's entire effort is a strategy to save as much of Cartesian realism as possible while giving up everything epistemological that would have returned us to the old Cartesian and Kantian paradoxes. So he is content to insert the parenthesis: 'Perhaps I should add that I think our actual scheme and language are best understood as extensional and naturalist.'[53] This is to warn us about the subversive possibilities of Kuhn's and Feyerabend's heterodoxy. But, clearly, is also signifies that, for Davidson, there must be a 'basic conceptual apparatus.' No wonder Davidson closes the argument in the following way:

> In giving up dependence on the concept of an uninterpreted reality, something outside all schemes and science, we do not relinquish the notion of objective truth – quite the contrary. Given the dogma of a dualism of scheme and reality, we get conceptual relativity, and truth relative to a scheme. Without the dogmas, this kind of relativity goes by the board. Of course truth of sentences remains relative to language, but that is as objective as can be. In giving up the dualism of scheme and world, we do not give up the world, but re-establish unmediated touch with the familiar objects whose antics make our sentences and opinions true or false.[54]

This is the Cartesian motive all right, but not its argument.

The essential issue is elusive to the last. The resolution of the incommensurabilism question, you realize, cannot be achieved by any merely 'factual' discoveries, as if by a patient review of the history of science itself, because the ultimate question asks whether *that* history should be read in a way that admits incommensurabilism from time to time or drums it out of play altogether. The question is insuperably philosophical.

Davidson had made a valiant attempt to support the second reading, but it backfired badly. If you ask what can be said in incommensurabilism's favor, the answer comes in three moves: the first (the bulk of what has been offered here) shows that the principal counter-strategies, those offered by Putnam and Davidson, fail hands down; the second shows the ubiquity of conceptual incommensurabilities *and* the reasonableness of subsuming epistemological incommensurabilities (the theme of incommensurabilism itself) among the common or garden varieties (even if rarer than most), and the third shows that if Cartesianism must be rejected (as it must) and the broad lines of Hegel's recovery of the grounds for a viable realism accepted (or, in effect, what realism must mean if it is to be recovered), then realism cannot be viable unless it is also a constructivism that is not foundationalist or transcendental at all, is historicized, or hospitable to relativism and incommensurabilism.

The upshot is that Kuhn and Feyerabend have hit on a genuinely fresh vein of inquiry that analytic philosophy has resisted (almost without compromise) through the whole of the last century. That may be the last outpost of Cartesianism. But if historicism can be revived, and that has yet to be seen, then English-language epistemology is likely to proceed along lines distinctly opposed to its own past history.

The curious fact remains: relativism, incommensurabilism, historicism, constructivism (in the epistemological sense) have been largely avoided in twentieth-century English-language philosophy – as much among the classic pragmatists and 'second-wave' pragmatists (Putnam, for example[55]) as among the analytically minded (Davidson, pre-eminently). This marks a deep gap in contemporary philosophy, which Kuhn and Feyerabend brought to professional attention in a compelling way. They have been drummed out of play, however, and essays like Davidson's 'On the Very Idea …' as well as Putnam's would-be (metaphysical) 'pluralism'[56] betray a deliberate restriction in addressing all the coherent options that are before us. The new century must understand that it is under an inherited obligation to make amends.

Notes

1 Putnam (1992), 'Wittgenstein on Reference and Relativism', p. 177; compare Putnam (1981), 'Two Conceptions of Rationality.'
2 See Burnyeat (1976).
3 See Winch (1958).
4 See von Ranke (1983).
5 See, further, Margolis (1991).
6 I discuss Putnam's account in greater detail in Margolis (2002a).
7 Putnam (1981), p. 216.
8 Davidson (1984), 'On the Very Idea of a Conceptual Scheme.'
9 Putnam (1992), p. 176.
10 Putnam (1987), pp. 17–18.
11 Putnam (1990), p. 120.
12 See Putnam (1987), Lecture II.
13 See Putnam (1999).
14 See Davidson (1984) and Putnam (1981).
15 See Dummett (1978).
16 Devitt (1996).
17 See, for instance, Kim (1996), p. 10.
18 See Davidson (1984), pp. 185–9. Compare Hacking (1982).
19 See Bernstein (1983); and Putnam (1981).
20 Davidson (1984), p. 190.
21 Ibid., p. 190.
22 I discuss this at greater length in Margolis (2003).
23 See Kuhn (1970a), Section X. See also Galison (1997), Ch. 1.
24 I have relied here on the research of Melissa Bowerman, part of which I have not seen in published form. A brief but convenient summary appears in Bowerman (1996), particularly pp. 337–46.
25 See Chomsky (2000). I explore Chomsky's new theory more at length in 'Materialism by Less than Adequate Means' (2002b).
26 Bowerman (1996), pp. 339–43.
27 See Wittgenstein (1969). I take this drift toward historicism to be part of the essential upshot of Wittgenstein's later philosophy.
28 Robert Brandom's recent 'inferentialism' is premised on a strong disjunction between sensory experience and enlanguaged belief or judgment, though it is also opposed (in favoring Wilfrid Sellars's well-known attack on empiricism) to the idea of the empirically 'given.' See Brandom (2000), Introduction and Ch. 1. You must also bear in mind that any such formula is profoundly anti-Darwinian. The argument – in Brandom, also in Richard Rorty – confuses or conflates the sense of 'sensation' or 'sensory experience' that is an ingredient in what is meant by 'sensory perception.' Where sensation and perception are distinguished in this way, 'sensation' designates a theoretically posited ingredient *in* perception: it is then never reported or reportable. The notion that sensory perception involves concepts, even among languaged animals (pigeons, for instance, trained to discriminate geometric forms), is perfectly straightforward. On the strength of the human paradigm, we regard the pertinent pecking behavior of the trained pigeon as manifesting conceptual competence of some kind. Concepts themselves are not empirically discerned, unless in a 'theory-laden' way. See Rorty (1998).

29 See Rosch (1978).
30 See Quine (1960), §§15–16.
31 Jerry Fodor springs to mind as a 'reluctant' Platonist of the innatist sort. See Fodor (1975). See also Katz (1966).
32 See Davidson (1986).
33 Feyerabend (1962), p. 82.
34 Davidson (1984), pp. 188–9.
35 Ibid., p. 188.
36 Feyerabend (1962), p. 89.
37 See note 2 above.
38 For the relevant background regarding Davidson's response to Kuhn and Feyerabend, see Davidson (1990). These are Davidson's Dewey Lectures. See also Davidson (1986) and Rorty (1986).
39 Compare Duhem (1962).
40 Kuhn (1970b), pp. 266–7.
41 Davidson (1984), p. 190.
42 Kuhn (2000), p. 266.
43 Ibid., p. 268.
44 Ibid., pp. 269–71.
45 Davidson (1984), pp. 184, 189–90.
46 See Whorf (1956). See also Quine (1953), 'Two Dogmas of Empiricism.'
47 Both Kuhn and Feyerabend would be unwilling to accept these simplifications. See Kuhn (2000) 'Commensurability, Comparability, Communicability' and 'Reflections on My Critics,' and Lakatos and Feyerabend (1999).
48 Giere (1988), pp. 36–7.
49 Ibid., p. 37.
50 There is an excellent appreciation of the matter in MacIntyre (1988). Unfortunately, MacIntyre solves the problem by a piece of conceptual machinery that violates the very condition he himself invokes in explaining its seriousness.
51 Davidson (1984), p. 183.
52 Ibid., p. 184.
53 Ibid., p. 188.
54 Ibid., p. 198.
55 See Putnam (1992).
56 See Putnam (1987), Lectures I–II.

References

Bernstein, Richard J. (1983), *Beyond Objectivism and Relativism: Science, Hermeneutics, and Praxis*, Philadelphia: University of Pennsylvania Press.

Bowerman, Melissa (1996), 'Learning a Semantic System: What Role do Cognitive Predispositions play?', in Paul Bloom (ed.), *Language Acquisition: Core Readings*, Cambridge, MA: MIT Press.

Brandom, Robert B. (2000), *Articulating Reasons: An Introduction to Inferentialism*, Cambridge, MA: Harvard University Press.

Burnyeat, Myles (1976), 'Protagoras and Self-refutation in Plato's Theaetetus', *Philosophical Review*, 85, pp. 44–69.

Chomsky, Noam (2000), *New Horizons in the Study of Language and Mind*, Cambridge: Cambridge University Press.

Davidson, Donald (1984 [1974]), *Inquiries into Truth and Interpretation*, Oxford: Clarendon.

Davidson, Donald (1986), 'A Coherence Theory of Truth and Knowledge', in Ernest Lepore (ed.), *Truth and Interpretation: Perspectives on the Philosophy of Donald Davidson*, Oxford: Basil Blackwell.

Davidson, Donald (1990), 'The Structure and Content of Truth', *Journal of Philosophy*, 87, pp. 279–328.

Devitt, Michael (1996), *Realism and Truth*, 2nd edn, Princeton, NJ: Princeton University Press.

Duhem, Pierre (1962) *The Aim and Structure of Physical Theory*, trans. (from 2nd edn) Philip P. Wiener, New York: Atheneum.

Dummett, Michael (1978), *Truth and Other Enigmas*, Cambridge, MA: Harvard University Press.

Feyerabend, Paul (1962), 'Explanation, Reduction, and Empiricism', in Herbert Feigl and Grover Maxwell (eds), *Scientific Explanation, Space and Time (Minnesota Studies in the Philosophy of Science*, Vol. 3, Minneapolis: University of Minnesota Press.

Fodor, Jerry A. (1975), *The Language of Thought*, New York: Thomas Y. Crowell.

Galison, Peter (1997), *Image and Logic: A Material Culture of Microphysics*, Chicago: University of Chicago Press.

Giere, Ronald N. (1988), *Explaining Science: A Cognitive Approach*, Chicago: University of Chicago Press.

Hacking, Ian (1982), 'Language, Truth and Reason', in Martin Hollis and Steven Lukes (eds), *Rationality and Relativism*, Cambridge, MA: MIT Press.

Katz, Jerrold J. (1966), *The Philosophy of Language*, New York: Harper and Row.

Kim, Jaegwon (1996), *Philosophy of Mind*, Boulder, CO: Westview Press.

Kuhn, Thomas S. (1970a), *The Structure of Scientific Revolutions*, enlarged 2nd edn, Chicago: University of Chicago Press.

Kuhn, Thomas S. (1970b), 'Reflections on My Critics', in Imre Lakatos and Alan Musgrave (eds), *Criticism and the Growth of Knowledge*, Cambridge: Cambridge University Press.

Kuhn, Thomas S. (2000), *The Road Since Structure: Philosophical Essays, 1970–1973, with an Autobiographical Interview*, ed. James Conant and John Haugeland, Chicago: University of Chicago Press.

Lakatos, Imré and Feyerabend, Paul (1999), *For and Against Method*, ed. Matteo Matterlini, Chicago: University of Chicago Press.

MacIntyre, Alisdair (1988), *Whose Justice? Which Rationality?*, Notre Dame, IN: Notre Dame University Press.

Margolis, Joseph (1991), *The Truth about Relativism*, Oxford: Basil Blackwell.

Margolis, Joseph (2002a), *Reinventing Pragmatism: American Philosophy at the End of the Twentieth Century*, Ithaca, NY: Cornell University Press.

Margolis, Joseph (2002b), 'Materialism by Less than Adequate Means', *Idealistic Studies*, 32, 3, pp. 273–90.

Margolis, Joseph (2003), *The Unraveling of Scientism: American Philosophy at the End of the Twentieth Century*, Ithaca, NY: Cornell University Press.

Putnam, Hilary (1981), *Reason, Truth and History*, Cambridge: Cambridge University Press.

Putnam, Hilary (1987), *The Many Faces of Realism*, La Salle, IL: Open Court.

Putnam, Hilary (1992), *Renewing Philosophy*, Cambridge, MA: Harvard University Press.

Putnam, Hilary (1999), *The Threefold Cord: Mind, Body, and World*, New York: Columbia University Press.

Quine, W.V.O. (1953), *From a Logical Point of View*, Cambridge, MA: Harvard University Press.

Quine, W.V.O. (1960), *Word and Object*, Cambridge, MA: MIT Press.

Rorty, Richard (1986), 'Pragmatism, Davidson and Truth', in Ernest Lepore (ed.), *Truth and Interpretation: Perspectives on the Philosophy of Donald Davidson*, Oxford: Basil Blackwell.

Rorty, Richard (1998), 'The Very Idea of Human Answerability to the World: John McDowell's Version of Empiricism', *Philosophical Papers*, Vol. 3, Cambridge: Cambridge University Press.

Rosch, Eleanor (1978), 'Principles of Categorization', in Eleanor Rosch and Barbara B. Lloyd (eds), *Cognition and Categorization*, Hillsdale, NJ: Lawrence Erlbaum.

Von Ranke, Leopold (1983), *The Theory and Practice of History*, trans. Wilma A. Iggers and Konrad von Moltke, ed. Georg G. Iggers and Konrad von Moltke, New York: Irvington.

Whorf, Benjamin Lee (1956), 'Language, Mind and Reality', in John B. Carroll (ed.), *Language, Thought, and Reality: Selected Writings of Benjamin Lee Whorf*, Cambridge, MA: MIT Press.

Winch, Peter (1958), *The Idea of a Social Science*, London: Routledge and Kegan Paul.

Wittgenstein, Ludwig (1969), *On Certainty*, trans. Denis Paul and G.E.M. Anscombe, ed. G.E.M. Anscombe and G.H. von Wright, Oxford: Basil Blackwell.

Index

'aboutness' relations 38–41
'absolute knowing' (*absolutes Wissen*) 86–8
accepted theories 141, 143; *see also soundly accepted* theories
Adam Bede 102
agency of science 140
Agincourt, Battle of 96
agrocentric systems 151–2
agroecological strategies (AES) 153–60
d'Alembert, Jean Le Rond 19–20
Althusser, Louis 63
Ampère, André Marie 20
analytic philosophy 32, 34, 41, 52, 67, 195, 206, 216
anomalous monism, theory of 39, 42, 45
anomalous phenomena 145–8, 154
anti-realism 38, 48–9, 54, 168, 181–2, 194–5
Aristotle and Aristotelian science 16, 46, 143, 146–7, 175, 183, 197, 207
astronomy *see* Copernican revolution
Augustine, St 106
autobiography 104–11

Bachelard, Gaston 34–5, 45–6, 50, 56, 59, 66–7
Barthes, Roland 63
Begriff concept 83, 88
Bergson, Henri 16
Bernoulli, Jacques and Jean 19
biodiversity 156
biotechnology 153–4, 158–60
Bloch, Marc 18
Bohr, Niels 47, 209
Boltzmann, Ludwig 9, 20
Borodino, Battle of 96
Boyle, Robert 176
Brown, Roger 102
Bruner, J.S. 106
Buridan, Jean 19
Burnyeat, Myles 208

calculus, differential and integral 20
Canguilhem, Georges 67, 179
Carnap, Rudolf 3, 16, 32, 36, 55, 180, 209
Carnot, Sadi 20
carriers for rules and norms 101–4
Cartesianism 3–6, 19–20, 32, 170, 194–209 *passim*, 213–16
Cartwright, N. 147
Cassirer, Ernst 2–3, 16, 181
causality, negative 116–17
charity, principle of 41–5
Chomsky, Noam 170, 202
Christianity 169
Clausius, Rudolph 9, 20
cognitive privilege 205
coherence theory of truth 118
Collingwood, R.G. 3, 132, 175
constructivism 3–5, 174–9, 182–4, 193, 195, 198, 209, 216
contextualism 37, 167–70, 183–4
continuity
  bearers of 100–104
  presupposed by change 99–100, 118
  theory of 122–9, 132–4
Copernican revolution 82, 169, 172, 175, 179–80
correspondence theory of truth 167, 181
critical philosophy 170, 175
cultural relativism 38, 41, 61
Curie, Pierre and Marie 11
custom 114

Danto, Arthur 132–3
Darwinian evolution theory 116, 170
Davidson, Donald 39–45, 52–4, 167–70, 177, 189–216
Dawkins, Richard 101, 117
deconstruction 51–2, 62–4
depth hermeneutics 41–2
Derrida, Jacques 34, 51–67
Descartes, René 1, 3, 10, 16, 20, 23, 31, 44, 194, 207, 213
Devitt, Michael 195

Dewey, John 194
Diana, Princess of Wales 101–2
'different worlds' doctrine 171–4, 177–8, 183–4, 196, 201
Dilthey, Wilhelm 80, 124, 132
double-aspect theory 39
Dummett, Michael 32, 48–9, 67, 194–5

Edinburgh group 179
efficacy of applied science 158
Einstein, Albert 9–16, 20, 23
Eliade, M. 125
Eliot, George 102
empiricism 38, 41, 52–3, 167–8, 211; *see also* logical empiricism
epistemology 31–2, 37–47, 51–4, 63, 167–70, 179, 192–8, 205, 209, 213–16
Euclid 59
Euler, Leonhard 19
Evans-Pritchard, E.E. 190
Evnine, Simon 42–3
evolution, biological 116–17, 170
experience, Kant's concept of 82–3
explanatory gap 180

factions 129
fame 101–2
farming systems, traditional 156–7
Feyerabend, Paul 2, 43, 52, 176, 189–216
'first person' narrative 108–11, 128
flags 101–2
'flashbulb memory' 102
Fleck, Ludwik 2
folk tales 105
food supplies 158–9
Foucault, Michel 63–5, 133
Frege, Gottlob 16, 31–2, 67
French Revolution 2, 6
Fresnel, Augustine Jean 9
'fruitful' strategies 145–50, 154, 159–61

Galileo (and Galilean science) 19, 143, 146, 169, 181, 197, 201, 207, 215
genetic modification 154
Géomètres, the 19
geometry 175
Gergen, K.J. 96
Gettsysburg Address 130–31
Giere, Ronald 212–13
globalization 159–60
Gödel, Kurt 57, 60–62, 66
grassroots organizations 152
gravitation, theory of 9
Green, T.H. 195
Grosseteste, Robert 19

Hamilton, Alexander 115
Hamilton, William 9
Hawkins, Harriet 105
Haym, Rudolph 80
Hebrew language 117
Hegel, G.W.F. 1–6, 53, 77–92 *passim*, 124–5, 171, 193–8, 216
Heidegger, Martin 35, 41–2, 128
Heine, Heinrich 80
Heisenberg's constant 47
Hempel, Carl 209
Herder, Johann 2
hermeneutics 41–5
heroic exemplars 114–15
Hertz, Heinrich 9
Hilbert spaces 21–2
historical construction 17–21, 116
historical narratives 99, 104, 112, 118, 121–7, 131–4
historical reality 127–8
historical view of knowledge 169, 177–8
historicism 5–6, 190–91, 194, 203, 216
historicity 3–4, 113–18, 124, 127, 200, 203–5, 209–11
  concept of 95–7
  definitions of 80–81
  denial of 139–40
  as the organization of facts 113
  problem of 98–9
  of science 4, 79, 82, 137–42, 148
history
  rejection of 126
  skepticism about 121–2
  as society's memory 128–31
Hobbes, Thomas 176
Homer 37
Hume, David 10, 36, 55
Husserl, Edmund 16, 31–5, 50–66 *passim*, 128, 179

idealism 132, 195, 198
incommensurability 171, 189–216
indexicality of narration 109–12, 118
inertia, principles of 19
insight, human 61
inten*s*ional as distict from inten*t*ional approaches 34–5
interactors 116–17
internal realism 54, 167, 177, 193
intuition 15–16, 22, 61, 190
intuitionism 49
irrationalism 182

James, William 180
Jefferson, Thomas 114–15
Julius Caesar 106

Kant, Immanuel (and Kantianism) 1–6, 10, 16, 20, 31, 36, 39, 44, 50, 55, 64, 77–92 *passim*, 170, 175–8, 182–3, 193–5, 198, 213, 215
Kelvin, Lord 9, 20
Kennedy, John 102
Kepler, Johannes 181
Kim, Jaegwon 195
King, Martin Luther 114
knowledge, theory of *see* epistemology
Koyré, Alexandre 19
Kripke, Saul 61
Kuhn, Thomas 2–5, 43–66, 137, 142–50, 154–5, 160, 169–84, 189–216
Kulke, Herman 125

Lacan, Jacques 63
Lagrange, Joseph 9, 19
Lakatos, Imre 50, 182–3
language, philosophy of 32
Laplace, Pierre Simon 9, 19
Latour, B. 113, 116
Lavoisier, A.L. 20, 196, 201, 205, 212–13
'laws of thought' 36
legends 131
legitimacy of applications 158
Leibnitz, Gottfried 16, 19, 36
Lévi-Strauss, Claude 63, 114, 125
Lewis, C.I. 3
Linacre College, Oxford 114
Lincoln, Abraham 114–15, 130–31
Locke, John 3, 10
logical empiricism 32–5, 44, 51–67 *passim*
logical positivism 2, 52, 57, 67
Lorentz, Konrad 9, 13
Luther, Martin 132–3
Lyotard, Jean-François 62

McClelland, D.C. 98
McDougal, W. 95
Marcus, Ruth Barcan 33–5
Margolis, J. 137
materialist strategies (MS) 143–59 *passim*
mathematics 9, 11, 17, 20, 23, 60–61, 90–92
Maxwell, James Clark 9
meaning-holism 35–6, 56
mechanics as analytic science 19
melody 99
memes 117
memory, social 128–31
Merleau-Ponty, Maurice 63
metaphysics 31, 66–7, 84–5, 142, 178, 198
Milgram, S. 96
mind-body dualism 39, 42
Mink, Louis 123
Mixtec languages 203
modern valuation of control (MVC) 151–4, 159–60
Mueller, Klaus E. 125
myths 131

Nagel, Ernest 209
narrative 104–6, 118
  artificiality of 123
  as *conjunctions* and as *disjunctions* 112–13
  conventions of 111
  and the creation of social time 113–15
  rejection of 126
  structure of 124
  see also 'first person' narrative; historical narratives; social narratives
narrative structure of human existence 125–8
naturalism 194
Neurath, Otto 3, 180
Newton, Isaac (and Newtonian science) 9, 23, 19–20, 133, 172, 175, 183

'noumenal' world 194
'now' as a relational expression 112–13, 118

'objective thinking' (*objektives Gedanke*) 87–8
objectivism 198–207
objectivity 10, 16, 178, 215
ontological relativity 49, 53, 65
ontology 45
Oresme, Nicole 19

paradigms and paradigm shifts 142–3, 171–6, 180–81, 197
parliamentary institutions 117
Parmenides 168–9
Peirce, C.S. 179, 183
Penrose, Roger 60–62, 66
Pericles 130
periodization 100
phenomenology 31, 34–5, 50–56, 62–7, 79, 85–8, 128
philosophy of science 2–5, 16, 31, 50–55 *passim*, 58, 63, 67, 116, 174, 179, 183
physical properties 7, 13–15, 19–21
physicalism 45
Plato (and Platonism) 52, 54, 167–8, 178, 190, 206
Poincaré, Raymond 9, 16, 20, 23
Popper, Karl 179, 182
positivism 52, 55–7, 66; *see also* logical positivism
posits 142–3, 184
post-analytic philosophy 35, 41–4, 61
postmodernism 62, 126
Priestley, Joseph 196, 201, 205, 212–13
propositional psychology 44–5
Propp, Vladimir 105
psychologism 51, 57, 59, 67
psychology *see* propositional psychology; social psychology
Putnam, Hilary 52–4, 167–8, 177, 184, 190–201, 216

quantum theory 8, 14, 20–23, 46–50 *passim*, 57, 170, 176
Quine, W.V. 32–46 *passim*, 49, 52–6, 65, 169, 179–80, 193, 200–201, 205, 211

Ranke, Leopold von 190–91
'rational contents' 9, 12
rationality 24, 77–84, 87–91, 139, 174, 182
realism 38–41, 44, 48–9, 54, 167–8, 174, 177–8, 183–4, 191–5, 198, 209, 215–16
  Cartesian 194
  strong form of 1, 3
  see also anti-realism; internal realism
reason, definitions of 77–8; *see also* rationality
re-enactments 132
Reformation, the 133
Reichenbach, Hans 2
relativism 17, 40, 43, 52–4, 63–6, 182, 184, 190–91, 196, 208–11, 214–16; *see also* cultural relativism
relativity, theory of 13–14, 20, 46, 170, 172, 183; *see also* ontological relativity
religious ceremonies 103
replicators 116–17
'revolutionary' science 45, 47, 59, 82, 138, 145–8, 183, 212
Ricoeur, Paul 63, 122
Riemann, Georg 20, 23
Rorty, Richard 38–54, 168
Rosenkranz, Karl 80
Ross, E.A. 95
Ruesen, Joern 122, 132
Russell, Bertrand 16, 32
Ryle, Gilbert 31, 67

Sachs, Oliver 111
Saussure, Ferdinand de 63
Schelling, F.W.J. von 195
Schrödinger's cat 47
Schwartz, Barry 115
scientific methodology 137–8, 141–3
scientific progress 174–6, 181–4
scientific revolutions *see* 'revolutionary' science
secularism 172
self, sense of 107–8
sensations 177
Shaffer, Simon 176
Shakespeare, William 96, 105–6
Shapin, Steven 176
Sherif, M. 96
slavery 131

social change 97, 115–17
social life 128, 134
social narratives 132–3
social psychology 95–6, 179
  absence of time in 98
  and historicity 118
  research design in 98–9
social time 113–15, 128
Socrates 208
solipsism 190–91, 196
soundly accepted theories 141, 144–7
story-lines 105
story-telling 107–11, 121–3, 134
strange attractors 21
structuralism 63–6
subjectivism 34
subjectivity 9–10, 14
supervenientism 196
sustainable ecosystems 155, 161
symbolic character of scientific knowledge 16–17

Ta'ziyeh tradition 103–4
theology 172
theory
  application of 149, 158, 161–2
  choice of 143
  derived directly from facts 180
  scientific 141
Thucydides 130
time
  everyday experience of 125–8
  nature of 111–15
  see also social time
Tolstoy, Leo 96
Toulmin, S. 116
tradition 113–14, 156–7
transcendentalism 3–5, 89
transgenic seeds 161
truth-claims 48, 60–65
truth-content 41–2, 58
truth-values 36, 40, 48

United States Constitution 130–31
universal grammar (UG) 202

value-outlooks 151–4, 157, 161
van Fraassen, Bas C. 184
verification-transcendent truths 48
Vico, Gianbattista 83, 89, 124
Vygotsky, L.S. 111

'warranted assertability' 48
Washington, George 114–15
Weber, Max 98
Webster, Daniel 131
Weinberg, Steven 182–3
Weyl, Hermann 20
Whewell, William 3
White, Hayden 123
Whorf, Benjamin 211
Williams, B. 168
Wills, Gary 130–31
Winch, Peter 190
Wittgenstein, Ludwig 35, 61–2, 99, 107, 111, 167–70, 183–4, 192, 203
Wolfson College, Oxford 114
world views 175–6, 179